Contents

	Page
Acknowledgements	iv

1. Introduction: Locality, rurality and social theory
    Tony Bradley and Philip Lowe 1

SECTION I   THE POLITICAL ECONOMY OF RURAL REGIONS

2. Rural regions in national and international
    economies   *Gareth Rees* 27

3. Capitalist restructuring, recomposition and the
    regions   *John Urry* 45

4. Segmentation in local labour markets
    Tony Bradley 65

5. The rural labour process: a case study of a
    Cornish town   *J. Herman Gilligan* 91

SECTION II   AGRICULTURE, LAND AND CAPITALIST PRODUCTION

6. Agrarian class structure and family farming
    Michael Winter

7. Land ownership and farm organisation in capitalist
    agriculture   *Terry Marsden* 129

8. Agricultural corporatism and conservation politics
    Graham Cox and Philip Lowe 147

9. The politics of rural landownership: institutional
    investors and the Northfield enquiry   *Richard Munton* 167

SECTION III   LOCALISM AND LOCAL PLANNING

10. The social meaning of localism   *Marilyn Strathern* 181

11. Women's roles and rural society   *Sue Stebbing* 199

12. The social effects of primary school closure
    Diana Forsythe 209

13. Images of place in a Northumbrian Dale
    Brendan Quayle 225

14. State planning and local needs   *Ian Gilder* 243

Notes on the Authors 258

Acknowledgements

We would like to thank members of the Rural Economy and
Society Study Group for their help and encouragement in
the preparation of this book; and the Economic and Social
Research Council for its financial support of the Study
Group and of a number of the research projects whose
findings are reported here.

LOCALITY AND RURALITY: ECONOMY AND SOCIETY IN RURAL REGIONS

edited by

Tony Bradley and Philip Lowe

published by: Geo Books,
Regency House,
34 Duke Street,
Norwich NR3 3AP,
England.

ISBN 0 86094 177 9

Printed at Short Run Press, UK.

1. Introduction: Locality, rurality and social theory

TONY BRADLEY & PHILIP LOWE

Massive upheavals in the national and international political economy are creating major changes in the economic systems, social relations and political balance of rural areas in Britain. Counterurbanisation, the decentralisation of certain forms of industrial activity and infrastructure, and the industrialisation of agriculture are powerful forces effecting rural social change. Their impact coincided in the 1970s with the extensive restructuring and centralisation of rural administration: including the reorganisation of local government; the 'regionalisation' of various statutory functions such as water and health provision; and the absorption of agricultural, fisheries and regional policy into the framework of the EEC. The attendant loss of local accountability has increased the political sensitivity of many rural issues. The flurry of policy making and investigation that has followed, has highlighted (as well as been handicapped by) our fundamentally inadequate grasp of the dynamic processes taking place within the rural economy and society.

To overcome this intellectual lacuna, the Rural Economy and Society Study Group was established in 1979 to provide a forum for all those studying the social formation of rural areas in advanced societies. It comprises social scientists from a variety of disciplines, including sociology, geography, economics, politics, anthropology and planning. Formed amidst a major renaissance of rural social research, the group has sought to encourage theoretically informed investigation and analysis of rural issues.

This book represents the first fruits of the collective effort of the study group, bringing together, as it does, many of those at the forefront of research activity. The focus of the book is on local life in contemporary rural Britain. Contributors were asked to reflect on recent empirical findings and theoretical developments. The papers thus contain a wealth of new material and analysis, combining various theoretical insights with detailed case studies ranging from the Scottish Highlands to Cornwall, from rural Wales to East Kent.

1

Some of the papers were first presented at a conference held in Oxford, in December 1982, convened by the study group and financed by the Economic and Social Research Council. In selecting papers for the book and asking authors to redraft their original contributions, the aim has been to produce a wide-ranging, yet rounded and integrated, volume representative of the breadth and vitality of current research. To this end, some papers have been omitted and others specially commissioned to close gaps in the coverage. Thus, the book should provide a unique point of reference for the burgeoning interest in rural issues.

In recent analyses of rural society there has been a reaction against the uncritical reliance of earlier studies on the overworked polarities of 'community-association', 'tradition-modernity', and the 'rural-urban continuum'. Investigators have reinvigorated the field by drawing upon challenging theoretical and methodological insights from other social sciences. This book refocusses attention onto social and geographical diversity by fusing together theoretical analysis from various critical perspectives with case studies of local life in contemporary rural Britain.

Two approaches run throughout the collection: firstly, a concern to understand and interpret the diversity of institutions and social relations evident in rural localities; and secondly, a theoretical grasp of how past and current restructurings of advanced capitalist societies have conditioned change in rural Britain. Within these broad parameters the perspectives adopted - concerning theory, methods and policy - are diverse but are linked together by three themes: the political economy of rural regions; the social structure and politics of capitalist-based agriculture and landownership; and government planning of rural community development. The many dimensions of the concept of 'locality' are explored, ranging from the ways in which people perceive and express their attachment to localities, to the ways in which localities respond to and are shaped by the decisions of government and the movements of industry, capital and labour over rural space. The contributors draw extensively on current research - their own and others. Our intention in this introduction is to relate the collection as a whole to past rural social research in Britain and to comment on its implications for future research directions.

THEORETICAL CONSIDERATIONS

Rural social research has no theory to call its own. That being said the keyword ideas on which its development rested until the 1970s - community, rurality, tradition, ascription, folk culture, dispersal - and the antonymic polarities to which it addressed itself - association, urbanism, modernity, achievement, mass society, concentration - were the bedrock concepts of early sociology and social planning. In conceptual terms, rural social research had become fossilised within a sociological time-capsule, anachronistically committed to a paradigm long since discarded empirically, as well as in social theory. These past failures aside, the important question is where can rural social research in Britain go in the immediate future?

This is not an easy question to answer at a time when, as Urry concludes (Chapter 3) sociology itself - especially 'urban' and 'industrial' - is in a state of flux. The common problem is one of the explanatory power of drawing connections between the artificial entities of 'geography' and 'sociology', of 'space' and 'society'. For, as Saunders comments, in summarising the complex sociology of knowledge contained in various perspectives on urban sociology:

> the basic stumbling block that all have
> encountered in different guises has
> been the need to relate certain social
> processes to particular spatial categories.
> The history of urban sociology, in other
> words, has been the history of a search
> for a sociological phenomenon the source
> of which may be located in the physical
> entity of the city.
> (Saunders 1981, p. 250)

It has been a long, often untidy and sometimes fruitless search. Nonetheless urban studies has been marked by its willingness, unlike its rural counterpart, to reject the ecological tradition, exemplified by the rural-urban continuum, and to move to formulate and test other concept-ualisations of the urban.

In contrast, the most successful recent attempts to integrate empirical rural research into mainstream debates within sociology and social history have shifted attention away from community studies towards an interest in the rural implications of *national* social change. Although perhaps an inevitable reaction, the eschewing of the community studies approach has had the negative effect of reducing the emphasis placed on the interpretation of *local* distinc-tiveness.

The death knell for traditional community studies and their imperative to delineate a distinctively rural way of life was sounded by Ray Pahl in 1966, when he demonstrated the spurious nature of the rural-urban continuum. Pahl concluded that "any attempt to tie patterns of social relationships to specific geographical milieux is a sing-ularly fruitless exercise". Instead he urged that socio-logical analysis in rural areas should focus on the contradiction between the local and the national, between the small-scale and the large-scale. Pahl's own research (1965) had been prompted by a rapid growth of residential inmigration into lowland villages. He perceived the potential implications of a trend to locate new development in commuter villages beyond the metropolitan green belt.

Theoretically he was concerned to investigate the effects of housing and planning policy on social cohesion, particularly for 'traditional' patterns of social relations based upon farming. Operating from a managerialist standpoint, he argued that the actions of local authority officers, particularly through the designation of land for private housing estates, were resulting in both the physical and social partitioning of villages. The dominant effect was irreversibly to alter long-standing patterns of spatial structuring and impose direct class conflicts onto village life. Pahl's initial thesis blazed a trail - he had struck a rich vein of popular discontent - to be followed by a new rash of largely defensive, anti-urbanism studies (e.g. Harris 1974; Ambrose 1974).

It was left to other writers to investigate the wider implications, in terms of political organisation, of 'counter-urbanisation'; including the emergence of environmentalist values (Lowe and Goyder 1983) and rural ideology, and their expression through local political coalitions and interest groupings within the various fractions of the middle classes (Newby *et al*. 1978, ch. 6). There remains a certain degree of disagreement, however, as to how popular anti-development sentiments should be theorised. On the one hand it is possible to see the mutual benefits which both farmers and amenity protectionists derive from restricted development, through either the maintenance of a low wage economy (Rose *et al*. 1976; Newby 1981) or the establishment of an exclusionary, private-sector housing market. On the other hand the attempt to theorise a political alliance between farmers and the new rural middle class makes little sense in terms of the conflicting goals of these 'property owners' (Buchanan 1982; Buller and Lowe 1982).

Curiously, few of these studies have had a *distinctively* local focus. Although the pioneering work in radical rural sociology of Howard Newby and his colleagues was locally grounded, its East Anglian setting was a distinctly subordinate feature, the region having been chosen because it approached an ideal-typical exemplification of the class-divided structure of British agricultural production, and of the position in the British class structure of farmers (Bell and Newby 1974) and agricultural workers (Newby 1977). Similarly, Lowe's researches into rural ecology and countryside protectionism have been part of a wider national investigation of environmental groups within the British political corpus, drawing upon contemporary theories of pressure group formation and organisation. A third strand of research, within the ambit of rural social geography, has been on national changes in rural housing markets (Dunn *et al*. 1981; Phillips and Williams 1982; Clark 1982; Shucksmith 1981; Winter 1980) and in the statutory provision and private distribution of services to relatively remote, sparsely populated areas (Moseley 1979; Cloke 1983; Mackay and Laing 1982; Moseley and Packman 1983). These geographical analyses of accessibility derive from spatial functionalism and ultimately central-place theory. As such they are directly descended from the ecological antecedents of the continuum idea. Although carried out within very different traditions, this range of studies share a common commitment to the examination of social changes in rural areas. The basic framing of research questions - using the formula, 'what have been the effects of national change X on the British rural population and class structure, as represented by area Y?' - has denied the specificity of local social processes and the social distinctiveness of localities.

THE LOCAL SOCIAL SYSTEM

Nevertheless, a variety of way-markers have been set in recent social theory, which return to the largely unresearched subject of identifying processes taking place in 'local social systems'. Ideas of community can be rendered down to three core themes: community as geographical locality; community as local social system; and community as communion, signifying common interest (Lee and Newby 1983, p. 57). From our standpoint it is the second definition which provokes the most interest. The first idea of community as a specific geographic area is spatial but not social and hence, from a sociological perspective, of only passing interest as a descriptive category. In contrast, the third notion of community, as a distinctive social category, has no necessary spatial significance. Nevertheless, by conjoining the spatial and social themes through the concept of 'local social system', it ought to be possible to examine the extent to which the social organisation of a particular locality demonstrates characteristics of localness and a sense of communion, or localistic commitment, amongst constituent social groups.

The idea of a local social system was originally suggested by Margaret Stacey (1969) in a paper outlining the myths to which community studies had been prone. Given its potential value, it is surprising how little it has been used by researchers (cf. Harris 1974). While the task of debunking the ideology of community has continued apace over the past two decades (Stein 1964; Gans 1967; Bell and Newby 1971), the substantive investigation of local social systems was eclipsed by these critical studies, largely on account of researchers confusing methodological with epistemological criticism. By eschewing the community studies method as a research strategy (as well as a mode of analysis), the baby was thrown out with the bath water, and rural research turned away from the interpretation of local distinctiveness.

With some writers, including Strathern (Chapter 10), having suggested that the 'local social system' is merely the shadow of community rediscovered, it is worth returning to Stacey's original paper to understand precisely what she meant by the term. To begin with, operating within a conventional Weberian framework of interpretation, she formulated the local social system as an ideal type - an analytical construct against which the social formation of actual localities can be compared. Its systemic components are social institutions: family structure, patterns of kinship, belief systems, political organisations, etc. A social system can only be said to be constituted locally when "a set of inter-relations exists in a geographically-defined locality". These inter-relations are unlikely to be complete, or totally confined within the boundary of the locality. Actual localities will display either partial or non-existent social systems. In consequence the model of a local social system should include an understanding of the spatial and temporal changes taking place to the social structure of specific localities.

In addition, Stacey considers a series of propositions useful for generating research hypotheses. In doing so, she presages much recent social theory in a manner particularly relevant to many of the contributions in this book. These propositions follow seven distinct themes: (i) temporal changes in the 'local' and 'newcomer' populations of the locality, (ii) convergence and overlapping of the social status and elite roles of local institutional actors, (iii) presence and absence of specific social institutions, (iv) articulation of local and national political structures, (v) strength of 'belongingness' and shared belief systems (communion), (vi) overlapping of local and national social systems, (vii) connection between geographical association and the strength of social networks.

Stacey suggests this as an empirical research agenda and a *modus operandi* for locality studies. In contrast to the community studies approach she intends such local analysis to make explicit the distinction between the content of social relations and the normative structures in which local social actors are embedded. In other words, she regards it as an open question whether, and to what extent, the social imagery of 'community' actually conforms to the structure of social practices. More than this, Stacey insists that the community concept is mythological.

The major advance implicit in Stacey's argument was for researchers interested in the local manifestations and extent of specific social institutions to study these freed of the obligation to make prescriptive observations about changes in the 'quality of life' and 'depth of human contact'. In this sense Stacey stands against the lengthy tradition of social philosophy concerned to demonstrate either the intrinsic 'moral affluence' of the rural or the 'moral degeneration' of the urban (see, *inter alia,* Williams 1973; Pahl 1967; Valentine 1968; Smith 1980; Saunders 1981; Donnison and Soto 1981, for critical discussions of this core theme of sociological theory). Nevertheless, despite the singularly fruitless nature of attempts to locate heaven in the countryside, hell in the city and a sort of purgatorial anomia in the suburbs these ideologies continue to inform everyday perceptions. Although 'way-of-life' studies are utterly discredited within the social sciences, these ideas – like those of 'deserving' and 'undeserving' poverty – remain as bulwarks of the wider society. They are, to borrow a metaphor from Golding and Middleton (1982), part of the morality play that is social policy.

The theoretical underpinnings of locality studies, as Stacey formulates them, are not without their own problems. The chief of these is the inevitable theoretical tension which afflicts all 'spatial sociology', of whether to locate social causation within local or non-local processes; and of how to conceptualise the connections between national and local structures. Writing in a former era of British sociology, heavily influenced by systems theory, her approach is blatantly functionalist. She explains the purpose of locality studies as a demonstration of the determinants of social integration. Consequently her most severe limitation lies in the complete neglect of the importance of local class formations in capitalist societies, and of social stratification in general. Her conception of locality studies never fully escapes from the dangers of theoretical 'holism', partly because of the undifferentiated manner in which she refers to the 'family' and 'belief systems' as institutions. Nevertheless, Stacey's paper is outstanding as an early attempt to come to terms with the social meaning of local diversity in the advanced societies.

To what extent has Stacey's local social systems agenda been implemented in rural research? As earlier studies and the various contributions to this volume show her proposals have received little attention. It is ironic that the most significant developments in the post-community studies era of rural research involved a retreat from the problematic social distinctiveness of localities into a search for the definitive characteristics of social relations established generally within rural areas.

Recently, though, there has been a strong revival of research interest in locality and localism from two quite distinct directions. The first, which derives from studies by social anthropologists of British rural communities, concentrates on the ethnography of locality, exploring the culture of localism and identification with place. The second, from within political economy, concentrates on the locality as the spatial focus for the reproduction of labour power within capitalist society. These two developments incorporate quite separate theoretical stances and method- ological emphases. Even so, both are important for the continued study of local social systems.

THE CULTURE OF LOCALISM

The recent growth of interest in the anthropological study of Britain has been assisted not only by the increasing financial and political inaccessibility of other, more exotic places; it has also been prompted by a populist tendency to cultural and political 'localism' which has

challenged the simplistic portrayal of Britain as a nation
having "a greater homogeneity by region and a greater
continuity over time than any other national society"
(Laslett in Rapoport *et al*. 1982, p. xii). It is ironic,
therefore, that localism and cultural diversity have become
a touchstone for the national organs of social integration -
the state, advertising and the mass media. Nevertheless,
this sense of local distinctiveness has provided an opportunity for social anthropologists to examine the phenomenon
of close attachment to locality nearer to home and to explore
the social meanings and expression of highly localised
collective identity within an advanced capitalist society.

    The volume of essays edited by Anthony Cohen (1982) is
representative of this work, most of which has focussed on
geographically peripheral communities; and the same author
has also drawn together a comprehensive review of recent
anthropological studies of rural Britain (Cohen 1983). The
new British ethnography owes a strong allegiance to the
phenomenological method - the meaning of people's social
experience is interpreted by recourse to the language,
values and beliefs through which they themselves depict
the everyday events of their social world. A common preoccupation is with identifying the symbols which structure
the forms of social life within particular localities.
As such, the shift of attention from exotic to indigenous
cultures not only implies differences in the style and the
circumstances of fieldwork but also affects the kinds of
study undertaken.

    As ethnographers of exotic cultures have frequently
stressed, field research relies on two quite separate
patterns of learning and modes of discourse (Powdermaker
1966; Burgess 1982). First, the researcher is required to
learn a new language in another culture. Second, the
professional anthropologist must preserve an outsider's
perspective and the capacity to distance himself or herself
from the new social situation. In so doing, "one becomes a
sort of double marginal man, alienated from both worlds"
(Evans-Pritchard 1973). But in the study of indigenous
cultures, the ethnographer, as a native speaker, is much
better equipped to penetrate the conceptual and idiomatic
subtleties of linguistic communities. This presents the
researcher with several methodological and conceptual
challenges.

    The task, according to Cohen (1983, p. 12), must be
"to penetrate the subterranean levels of meaning: to explore
not generality and the norm, but to seek out *diversity,* to
explore differences *within* the culture - to see, therefore,
not how culture *integrates* but how it *aggregates*". This
calls for the researcher to go beyond the observation of
linguistic equivalence and usage towards a perception of
the overlapping structure and ambivalence of cultural forms.
A primary concern must be the manner in which social boundaries are generated and maintained. "The ethnography of
locality" writes Cohen (1982, p. 2), "is an account of how
people experience and express their difference from others,

and of how their sense of difference becomes incorporated
into and informs the nature of their social organisation
and process. The sense of difference thus lies at the heart
of people's awareness of their culture" (cf. Barth 1969;
Strathern, Chapter 10).

This approach predisposes investigators to analyse
particular places as boundaries of commonality. Unlike an
earlier generation of community studies, however, there is
much less of a tendency to treat geographically peripheral
communities as isolated from the wider socio-cultural
milieux. On the contrary, the local implications of changes
in the complex economic and political systems of Western
Europe have provided a primary focus for research. Even
so, the price of cultural familiarity is the possible
blunting of the critical faculty of self-awareness, through
an all-too-easy identification with local social actors.

A more serious weakness of British ethnography, despite
its current signs of vitality, is the absence of a strong
research tradition. As students of their own society,
British anthropologists compare unfavourably with their
counterparts in North America and in many other European
countries. They cannot, from indigenous sources, draw upon
an accumulation of field experience or a mature debate
concerning theory and methodology (Cohen 1983). One consequence is an overemphasis on describing and interpreting
the particularities of local idioms and social forms to the
neglect of comparative analysis and abstraction. The danger
is that the field may lapse into the sterile parochialism
and uncritical acceptance of the rural idyll which plagued
some of the earlier community studies.

These pitfalls are carefully avoided in the contributions
to the new British ethnography included in this volume by
Gilligan (Chapter 5), Strathern (Chapter 10), Forsythe
(Chapter 12) and Quayle (Chapter 13). Indeed, they address
issues and concepts which should help to provide a firm
basis for the development of a vigorous and cumulative
research tradition, such as the analysis of changing economic
patterns, the interpretation of symbolic forms, the meanings
of localistic attachments, the experience of marginality
and the articulation of local interests.

RURALITY AND REGIONALISM

Whereas the primary concern of the new British ethnography
is with the cultural manifestations of localism, an alternative perspective for theorising the social construction of
localities arises from work on the political economy of
regional development and underdevelopment, particularly the
rethinking prompted by changes in the location of manufacturing employment, including the out-movement of industry
into peripheral and rural localities. Much of current
theorising is rooted in critiques of Marxian dependency
theories.

Although there are considerable variations in detail
and approach, dependency theorists maintain that the divisions
and relationships between regions are a function of the
process of surplus extraction (see, *inter alia,* Amin 1976;
Emmanuel 1972; Frank 1979; Wallerstein 1974). In these
terms, the decentralisation of manufacturing industry is
viewed as one stage in the continuous reproduction of an
unequal relationship between peripheral and core regions,
wherein the latter extracts surplus value from the former.
At first, the source of surplus value is 'peasant' or
'simple commodity production' in agriculture. This is
succeeded by the employment of the peripheral labour force
in branches of manufacturing industry (Lipietz 1977) and,
finally, the development of tertiary sector production in
the periphery (Lipietz 1980).

The many varieties of dependency, core-periphery and
unequal exchange theories have been trenchantly attacked on
a number of grounds (see Cooke 1983). They have been
criticised for a static and ahistorical conception of the
processes of development (O'Brien 1975). Moreover, in
focussing on relations of exchange and processes of circulation, they overlook the social and economic relations
established at the point of production, through the labour
process. Of direct concern here, as Winter demonstrates
in his critical evaluation of theories explaining the
'survival' of family farming (Chapter 6), dependency theories
continually encounter the problem of how to conceptualise
the incorporation of 'peasant', 'precapitalist', 'non-
capitalist' or 'feudal' social relations into capitalist
relations of production. As recent conceptualisations of
labour market stratification emphasise (Bradley, Chapter 4),
capitalist society is structured around the subjection of
labour socially as well as in legal form (Corrigan 1977),
in all localities, be they rural or urban. Consequently,
to build social theories on the depiction of particular
spaces - localities, regions, nations - as generically
backward, underdeveloped or permanently peripheral ignores
the dynamism of spatial and temporal change in the means of
exploitation, within capitalist relations of production.
Indeed, the very idea of a 'region' as a distinctive socio-
spatial entity has been critically appraised for both the
blunt acceptance of conventional state categories and the
consequent confusions to which 'regional analysis' is prone
in its elision of highly complex social and economic changes
with a simplistic spatial classification (Massey 1979;
Markusen 1980; Sayer 1979; Pickvance 1983).

A number of more sophisticated theories have been
developed from the above criticisms of notions of dependent
development. In Britain this has been marked by a vigorous
attempt to explain the processes whereby capitalist accumulation alters the spatial structure of the society. These
theories of capitalist recombination - as exemplified in
the writings of Philip Cooke (1980, 1982, 1983), Doreen
Massey (1981, 1982) and John Urry (Chapter 3) - specifically
address the question of industrial decentralisation. Though
it is not possible to do justice to the richness of the

arguments, for our purposes we regard the main points of
the approach to be the following.

The drive for capitalist accumulation generates changes
in the relations of production such that capital switches
attention from issues simply to do with labour productivity
to those of more efficient circulation. In its monopoly
phase capital has become *relatively* detached from spatial
constraints. Industrial capital is, therefore, spatially
indifferent and can relocate to make use of cheap and
compliant labour reserves. Capital restructuring alters
and reconstitutes various technical divisions of labour
within industry, thus creating new spatial divisions of
labour, particularly whereby higher and lower level functions
of specific firms and enterprises are spatially split,
for example through the location of branch plants in peripheral regions. Productive decentralisation into the
periphery may arise from a variety of causes - such as
rationalisation, or the reinvestment of fixed capital -
rather than as the ineluctable outcome of any particular
logic of capitalist development, beyond the impulse to
increase the value of labour exploitation.

In contrast to the increasing mobility of capital,
labour remains relatively immobile. Thus, localities -
primarily cities - have become sites wherein labour power
is produced and reproduced within households, through the
consumption of commodities. The recomposition effect of
capitalist restructuring entails the production of local
uniqueness, whereby particular places come to display the
characteristics of specific class and social structures.
Particular localities and 'types' of locality - such as
inner cities, or sparsely populated rural areas - may come
to be regarded as spatially and socially disadvantaged on
account of local stratification structures. Nevertheless,
no local social structure is the product of one single
spatial division of labour. Rather, each locality is
produced by the complex *recombination* of antagonistic class
interests, which result from successive rounds of capital
circulation, the overlapping of industrial tiers and labour
restructuring within the locality.

We have summarised these arguments in some detail
because of the importance that they have for establishing
a research agenda for the examination of rural localities.
Even so, certain difficulties, particularly concerning
actions within the working class, attend many of the issues
raised. For example, there is a tendency to reduce *social*
change to the effects of capital penetration of localities,
with the implication that the local workforce is compliant
to the demands of new industrial capital.

A more general point, which relates to Urry's specification of 'local stratification/social class structures',
is that the social relations established at a local level
are theorised as arising from the combination of national
class groupings, under the circumstances generated by

11

overlapping spatial divisions of labour. Despite Urry's
emphasis on the importance of examining *local* social
relations these seem to consist of those classes within
the 'national' configuration of the capitalist mode of
production which happen to be represented locally. An
alternative view would emphasise the need to conceptualise
the 'new' local processes of stratification generated
through recombination effects. A related point is the
question of the formation of local political alliances to
attract or oppose new capital penetration. Recently,
Pickvance (1983) has suggested the term spatial coalition
- or local coalition - to cover the various groupings of
capital fractions, labour strata and local states responding
to capital restructuring. His argument is that we need
more 'botton-up' studies of local action to complement
'top-down' structural accounts.

What is underdeveloped in these approaches, arising
from the emphasis placed on examining the geography of
capital recombination, is an understanding of the dynamic
processes affecting the stratification of working class
groups in peripheral localities. In this respect, further
work is required elaborating the discontinuities between
labour markets which serve to segregate the working class
spatially. Despite the many and considerable differences
between approaches, recent research within the new British
ethnography, commented upon above, may provide the starting
point, particularly in interpreting the significance of
local identification for defences against or resistance to
the imposition of new labour processes. It is at least as
open a question what impact localism has on industrial
production relations as the impact that new industrial
branch plants have on social relations established within
rural localities.

Moreover important changes are taking place involving
the integration of small businesses, both within and outside
agriculture, into capital monopolies. Consequently, research
is required into the precise strategies and relationships
developed between the various local petty bourgeoisie and
'newcomer' capitals, particularly as guided by the local
state's advocacy of alternative land use and economic
planning policies. Such research would, in turn, need to
address issues concerning the state's involvement in social
reproduction in rural localities, through investment in
productive infra-structure (roads, advanced factories,
sewage and water systems), social and collective consumption
(schools, housing, community centres) and directly through
the 'social costs' of local government (planning, administration, development agencies).

The three perspectives outlined here - the sociology
of localities as 'local social systems', the ethnography
of localism, and the political economy of capitalist
recombination in peripheral regions - all contribute
valuable insights to the examination of social processes
taking place within rural localities. They also decree

certain imperatives for the social researcher wishing to investigate these phenomena. First, the research must be thoroughly grounded in detailed empirical work within specific localities. Second, the researcher needs a thorough understanding of the relevant social theory. Third, the research agenda developed will need to be comparative, not merely so as to examine the differential effects of 'national change X' on 'localities A, B, C', but in order to specify the unique configuration of the local 'civil society'. The various themes developed in the contributions to this volume provide a starting point for such a comparative research agenda.

The exuberance and vitality with which rural social research is currently invested have meant that particular objects of research have been studied from very diverse perspectives, ranging from the methodological individualism of much social anthropology to the structuralism of much political economy. Indeed, there are perhaps greater contradictions between the approaches adopted in the following chapters than there are points of consensus. Given the 'totality' of topics covered this is as it should be. Indeed, we regard it more as a strength than a weakness. One point, nevertheless, needs to be emphasized. As should be clear from this introduction and what follows, there is no heuristic value in the 'rural' epithet. It is at best a convenient conventional label to attach to social research carried out within particular geographical contexts. Localism and the recombinative effects of contemporary capitalist development may be examined in any locality. That being said, the precise implications of such changes in specific 'rural' localities indicate a number of the most important and significant changes currently taking place in British society. For this reason the studies that follow, by returning us in more coherent fashion to many of the issues raised by Stacey's conception of the 'local social system', are valuable as much for the questions that they raise as those that are answered.

THE CONTRIBUTIONS

The contributions are grouped into three sections: the first on the political economy of rural regions; the second on agriculture, land and capitalist production; and the third on localism and local planning.

In the first contribution to Section I, Gareth Rees examines the spatially shifting nature of recent capitalist restructuring with reference to the political economy of rural Wales (Chapter 2). His analysis proceeds from a critical appraisal of the central feature of radical rural sociology, namely, the fundamental importance of land and property relations in conditioning rural social change. Conversely, Rees argues that regional development is more systematically conceived as the result of changes in the location of fractions of capital, searching out ever more advantageous industrial sectors and spaces to expand into. Consequently, rural localities are produced and their endemic social relations reproduced through the logical workings of *uneven* capitalist expansion. As such, it is misleading to focus attention onto agriculture *per se*, which represents merely the dominant capital fraction of a preceding round of investment and capital circulation. Rees elaborates this theoretical position by reference to productive decentralisation in rural Wales.

Many of the issues raised by Rees are expanded upon theoretically by John Urry, one of the leading writers on capitalist recombination, in Chapter 3. Here he develops a discussion of the nature of civil society and local class structure, raised in earlier work, and applied to the specification of 'rural' localities. His method is to subject the core ideas of spatial and, particularly, rural sociology to trenchant criticism. He argues that the 'rural' is a descriptive category lacking any explanatory power. Rather, he demonstrates that alterations to the forms of capitalist and state restructuring act to cause the production of specific local economies and social structures. As with Rees, he identifies the effects on rural labour of capitalist restructuring as of crucial significance. Localities serve as labour pools and specific places will differentially attract or repel capital in search of suitably productive workers. Thus localities develop particular patterns of stratification, which condition whether the actions of social groups display greater or lesser degrees of class solidarity. Various features of civil society developing in rural localities serve to reduce the saliency of class divisions. The implications of Urry's analysis is that political action in rural localities becomes largely confined to issues of collective consumption, the maintenance of unequal access to property and the social segregation of space.

The development of specific local labour markets is the subject of the remaining two papers in this section. In Chapter 4 Tony Bradley analyses empirical data from a large-scale survey of households resident in five quite different English localities, in order to establish the degree to which rural labour is segregated into specific labour markets. Drawing upon recent theories of labour market differentiation he examines certain key variables which embue individuals with greater or lesser capacity to gain access to a variety of employment opportunities. Bradley's intention is to comment upon the hypothetical causes of the operation of manifest asymmetries of power dividing the rural working class. In addition to the use of exclusion strategies arising from the deployment of educational qualifications and professional credentials, he suggests that the degree to which rural labour is attached to localistic networks, through which jobs may be allocated, is a prime determinant of who gets which jobs, where and how. Localism will, therefore, act as a mechanism whereby inequalities of employment opportunities develop as sources

of social stratification for rural labour. Within the
primary asymmetry dividing labour from capital further
divisions exist: generally speaking, women, locals and the
lower class fare worse in the labour market than do men,
newcomers and the upper middle class.

Herman Gilligan's contribution (Chapter 5) puts some
descriptive flesh on the analyses presented in the previous
chapters. He presents an ethnographic case-study of the
impact of historical and contemporary economic changes on
the small Cornish town of Padstow. The effects of successive
rounds of capital investment in Padstow has produced a
relatively undifferentiated working class, with historical
roots in fishing, shipbuilding and small business. Of
particular importance has been the development of extensive
networks of informal contacts and localistic ties between
small capital owners and local labour, largely as a result
of the marginalisation of the local working class following
the decline of the indigenous shipbuilding industry.
Gilligan describes the differential strategies available
to local labour, through the use of property and kinship
ties, and other local solidaristic relationships, which
have created and sustained labour market and income opportunities in both the formal and informal economic sectors.
The homogenisation and marginality of labour ensured a
ready acceptance of more stable work conditions in new
manufacturing industry, when such capital arrived, in the
form of a branch of a London-based printing company. The
impact of new industrialisation was to fragment the local
working class and sever many of the existing localistic
ties. The question is raised of what will be the longer-
term effects on stratification and geographical and social
mobility. Clearly, a rich research agenda follows from
this and the other papers in the section, for the invesigation
of rural labour markets.

The four contributions in the second section address
the theme of the involvement of the state and capital in
the conditions of agricultural production and the circulation
of land uses in the British countryside. These two issues
have been central problems for radical rural sociology.
Each of the new essays develops the study of the ownership
and control of land qua property relations through examining
the specific characteristics of social and political relations
manifest in contemporary agricultural change.

The question of how family farming has survived throughout the era of monopoly capitalism is taken up by Michael
Winter (Chapter 6). In so doing, he returns us to the
theme of how to conceptualise the uneven development
processes in the advanced societies. His argument is based
on the fact that as a productive category the fundamental
class division between labour and capital is absent from
family farm households. Nevertheless, he rejects the
various theoretical traditions which identify kinship and
family characteristics as the determinants of family-based
production. As he emphasises, family farm enterprises exist

to further accumulation, albeit of a small-scale nature, and not merely the consumption and social reproductive needs of the household unit. Winter examines the proposition that the peculiar nature of land as a means of production in agriculture and the resultant natural monopoly conferred on any occupier have acted to impede the penetration of outside capital and to facilitate the resurgence of family farming. The rent relation and the relative length of the productive cycle further distinguish agriculture from other forms of commodity production. He remains critical of functionalist explanations that characterise contemporary family farming either as a historical point of transition or as a source of value for monopoly capitalists. It is unlikely, Winter suggests, that any single theory can provide a thoroughgoing explanation of the relation of family farming to capitalist production.

In concluding his chapter Winter stresses the importance of examining the internal characteristics of family farming, especially the accumulative pressures for expansion which may signal its denouement as a unique productive form. Terry Marsden provides such an analysis in Chapter 7, which elucidates the peculiarities of capital accumulation and concentration that have occurred in British agriculture over the last hundred years. Drawing on his investigations of arable farming in eastern England, Marsden describes how these processes have led to a restructuring of landownership and holding, with the consequent emergence of complex and large-scale farm businesses, dominating agricultural production and integrating it into the circulation of finance and other sectoral capital fractions. Marsden shares Winter's concern with the productive significance of land, although he stresses the tendency of farming capitalists to treat property as a prime source of finance. Even so, Marsden argues that family farming is not so much a peripheral as the primary social form of capitalist agriculture. Unlike the family farm in Victorian England contemporary owner-occupier households are now more effective in reproducing an appropriate labour force. Thus, kinship and family characteristics continue to play an important part in determining the precise structure of business, landholding and labour organisation, facilitating the expansion of capitalist relations of production in agriculture. Furthermore, these social relations are entirely compatible with the integration of farming into other systems of production which have no such reliance on the use of land.

The other two chapters in this section take up the ideological and political dimensions of agricultural production and rural landownership, with particular attention to the role of dominant rural interests in determining state policy. Graham Cox and Philip Lowe (Chapter 8) examine the farming lobby's close relationship with the state which gives it such a strategic advantage in dealing with opposing interests, such as the rural conservation movement. The relative power of these two lobbies is compared in the

framing of the Wildlife and Countryside Act 1981. It is
suggested that the established relationship between agricultural interests and government made available to policy
makers a corporatist policy option which depended on the
farming community policing its own activities, and thereby
retaining its autonomy and freedom from statutory controls.
The authors conclude that this special relationship partly
reflects the power of the National Farmers' Union and the
Country Landowners' Association, but derives ultimately
from government commitment to the agricultural sector and
a particular (corporatist) mode of state intervention,
based on perceptions of its success.

The subject of Richard Munton's contribution (Chapter 9)
is the political controversy over changes in land tenure,
particularly the purchase of farmland by the financial
institutions, and the handling of the issue by the government
and the agricultural lobby. He shows how political concern
over rising farmland prices and a decline in opportunities
to enter farming became focussed in the late 1970s on the
financial institutions' growing involvement in the rural
land market. To many observers, institutional investment
in agricultural land seemed singularly inappropriate because
of its traditional use as a primary resource by private,
family enterprise. An official Committee of Inquiry,
chaired by Lord Northfield, was set up in 1977 to investigate
the issue. Munton shows how eventually the Committee and
indeed the farming lobby came round to a position of
tacitly endorsing the financial institutions when it became
clear that any restraints on their activities would entail
restrictions on the free market in land; and that they were
the only source of expanding landlordism. The incident
reveals some of the contradictions in the ideology of
British farming, with the imperative of an unfettered land
market and the myth of the farming ladder prevailing in this
instance over the potent imagery of the owner-occupied
family farm.

The third and final section provides a variety of
insights into the importance of localism and rural culture
as expressions of ideologies which inform everyday life
in British villages. These serve to differentiate the
content of social relationships established in country
localities from those pertaining in towns. As such, they
are instrumental in conditioning the form of popular
responses to decisions affecting collective consumption and
social reproduction taken beyond the locality by the state
and capital. In the three papers by Diana Forsythe,
Brendan Quayle and Ian Gilder specific aspects are covered
of the contradictory processes of state planning whereby
the state seeks both to facilitate accumulation by reducing
social expenses and to provide for the continuation of
local civil society. To borrow a distinction used by Goffman,
in a not entirely different context, these papers move us
from the 'front region' of the political economy into the
'back region' of the rural localities, wherein the social
formation is lived and experienced.

In the first of these studies Marilyn Strathern
(Chapter 10) considers the social meanings of 'localism'
by returning us, in a wholly original way, to one of the
earliest themes of rural research, namely, the mythology
attaching to ideas of the village. Strathern describes
the village as not only, or primarily, a particular place
but rather as a concept for which there is no fixed social
or geographical referent. Who, where or what the village
is depends on who you are: your gender, your occupation,
your kinship links locally and beyond, and whether your
connections with a place are through birth, marriage,
residence or work. She draws on ethnographic research
carried out in one Essex village, by several generations of
Cambridge anthropologists, to adduce the meanings of what
it is to be a 'real, local Elmdoner'. As she explains,
ideas about 'localism', as well as definitions of who is
local, are essentially contestable idioms and values, used
to serve specific interests. Ultimately they reflect ideas
about class formation, gender divisions and social mobility
in English society *tout court*. The peculiarity of English
notions of class is the apparent contradiction between
fixed strata and mobile individuals. This is manifest in
Elmdon by the extent to which privacy, and not community,
is defended so as to facilitate mobility (social or geographical), whilst the bounded category of the village
remains, as if immutable. Such contestable idioms are
also evident in the diversity of analytical and prescriptive
categories that social scientists employ in describing
localism and class relations. Indeed, Strathern's paper
may be read partly as the description of a particular local
social context, partly as a discourse on the nature of the
class formation in English society, but also as a way of
reading the social theories employed as analytical strategies
throughout this collection.

As pointed out above, rural community studies have
continually returned to the impact - usually labelled
social polarisation - of a new, residential, middle class
on the local social structure. This approach is best
represented in the present volume by the paper from Susan
Stebbing (Chapter 11), in which she examines the differentiation of women's status roles, according to their identification with the idea of rootedness in the countryside and
its attendant traditions. She reports on an extended
study of the women residents of two parishes in rural Kent.
Stebbing classified her respondents as having either
'traditional' or 'non-traditional' sex roles, according to
whether they viewed themselves as home-centred, nurturant
and subordinate or, in contrast, espoused values of sexual
equality, job-centredness, freedom and self-fulfillment.
As she convincingly demonstrates, the image of "being a
country woman" provided the majority of her respondents
with an essential means of employing attachment to the
locality as a reinforcement of the stereotype that "a
woman's *place* is in the home". In tracing the ideological
roots of this imagery to the women's reference groups
Stebbing draws attention to an important feature of locality

studies identified by Stacey, namely the examination of social networks. Essentially, the countrywomen maintained predominately horizontal contacts within the locality, whereas the less traditional women were able to hold onto their own self images largely through involvement in networks established beyond the local social system.

Certain severe problems of interpretation arise from adopting a solely political economy perspective in studying rural social change. The value of alternative approaches is highlighted by the various papers in this final section and none better than in Diana Forsythe's analysis of the social effects of the closure of village primary schools in Scotland (Chapter 12). When faced with the suggestion that a school should be closed, politicians, teachers, local government officials and local people, all use similar language, describing the prospect essentially as a violation of the local community, invoking the imagery of death and destruction. From this basis Forsythe sets out systematically to describe the actual experiences of local people when faced with such a potential violation. She found that, at least within Scottish culture, educationalists and villagers share the same perceived *functions* of the school: as providing for the education of children, rather than as a resource for community development. Nevertheless, when threatened by closure local people emphasise the *social* function of the schools. Herein, Forsythe refreshingly knocks a number of community development myths on the head. The school is not particularly important as a centre or as a source of local leadership. It has greater significance in the opportunities that it provides for social interaction, integration and as a representative symbol of local identity. When closure has taken place, the children have often benefitted educationally but the community has lost a crucial facet of its social identity.

This theme - the identification and idealisation of place - is taken up by Brendan Quayle (Chapter 13) in his ethnographic study of community and landscape symbolism in Allendale, Northumberland. Quayle contrasts two contexts of change affecting this upland locality, one involving 'development' and the other epitomising 'decline'. An example of the former was the threatened conversion to a ski centre of the former school house in the village of Sinderhope. As he portrays it, Sinderhope has no 'natural' topographical focus. In consequence, the school house held an especial significance for the residents as a symbolic centre. Its threatened fate - to be used and controlled by outsiders - mirrored the wider threat facing the whole community. Local opposition was intense. In order to restore the school house as a village institution and to reassert local control, the residents mounted a campaign for it to become a community centre. Resentment over the pressures from 'outsiders' and 'newcomers' on local services and housing is further demonstrated in local support for new industry, and in opposition to various conservation proposals for the area as being out of keeping with its industrial heritage. In contrast, new rounds of investment and the circulation of capital into the area to exploit tourist services are seen by locals as signals of decline and decay, whereas in Gilligan's Padstow they were viewed with a certain optimism as providing welcome employment opportunities. The protection of the school at Sinderhope (for how long?) is presented as a source of some optimism in distinct contrast to the despondency, portrayed in vivid detail, of the Allenheads population, who perceive their place as decaying in the neglectful hands of the once nurturant local estate and returning to the chaos of the wild fell. Quayle emphasises, in his account, the homology between the decline of the place and the neglect of the people which the residents feel. Nevertheless, he closes on a note of hope, by pointing to the integration of 'newcomers' into the local society or, at least, of those who share the ethnic locals respect for their place.

In the final paper of the collection (Chapter 14) Ian Gilder, a senior rural planner, addresses some of the pressing problems which confront the state in adequately responding to the needs of the local people. His theme is that increasing centralisation of policy making, despite the political saliency of localism, has reduced the potential for local authorities to intervene in providing for the real needs of localities. Moreover, as he outlines, the vehicle for mediating central state policy to local government – land-use development planning – by giving aspatial policies a spatial perspective, is singularly ill-equipped to respond to changing local *social* needs. Rural planning policy has been founded on an overriding concern to contain 'urban' development, although new initiatives to attract development into peripheral localities may be altering this single-minded devotion to rural preservation. Gilder examines the effectiveness of policies explicitly designed to meet local needs with specific reference to the provision of public sector housing and the apparently more radical but, in reality, toothless powers appropriated by certain local authorities to intervene in the sale of newly built private housing. In both these policy arenas, the record of local planning authorities has been pitifully poor and inequitable. Significantly, Gilder points to the work of social anthropologists as providing a crucial lead in explaining the meaning of local need to statutory authorities. Even so, the evident powerlessness of rural people to influence their own planned destiny (or deprivation) – a theme running throughout this collection – is exacerbated by political neglect. In conclusion, Gilder suggests three directions of change towards increased local planning autonomy and the redistribution by the state of the social product to rural localities. Without such changes – which, in the current context of the political economy of central-local state relations, do not seem to be close at hand – the prospects for rural localities appear bleak.

REFERENCES

Ambrose, P., 1974. *The Quiet Revolution,* (London: Chatto & Windus).

Amin, S., 1976. *Unequal Development,* (Brighton: Harvester).

Barth, F., 1969. *Ethnic Groups and Boundaries,* (London: Allen & Unwin).

Bell, C. and Newby, H., 1971. *Community Studies,* (London: Allen & Unwin).

Bell, C. and Newby, H., 1974. Capitalist farmers in the British class structure, *Sociologia Ruralis,* 14, 86-107.

Buchanan, S., 1982. Power and planning in rural areas, in: *Power, Planning and People in Rural East Anglia,* ed M. Moseley, (Norwich: Centre for East Anglian Studies).

Buller, H. and Lowe, P.D., 1982. Politics and class in rural preservation, in: *Power, Planning and People in Rural East Anglia,* ed M. Moseley, (Norwich: Centre for East Anglian Studies).

Burgess, R.G. (ed), 1982. *Field Research: A Sourcebook and Fieldmanual,* (London: Allen & Unwin).

Clark, G., 1982. *Housing and Planning in the Countryside,* (Chichester: Research Studies Press).

Cloke, P., 1983. *An Introduction to Rural Settlement Planning,* (London: Methuen).

Cohen, A.P. (ed), 1982. *Belonging,* (Manchester: Manchester University Press).

Cohen, A.P., 1983. *Anthropological Studies of Rural Britain 1968-1983,* (London: Social Science Research Council).

Cooke, P., 1980. Dependent development in UK regions with particular reference to Wales, *Progress in Planning,* 15, 1-62.

Cooke, P., 1982. Class interests, regional restructuring and state formation in Wales, *International Journal of Urban and Regional Research,* 6, 187-203.

Cooke, P., 1983. *Theories of Planning and Spatial Development,* (London: Hutchinson).

Corrigan, P., 1977. Feudal relics on capitalist monuments? Notes on the sociology of un-free labour, *Sociology,* 11(3).

Donnison, D. and Soto, P., 1981. *The Good City: A study of Urban Development and Policy in Britain,* (London: Heinemann).

Dunn, M., Rawson, M. and Rogers, A., 1981. *Rural Housing: Competition and Choice,* (London: Allen & Unwin).

21

Emmanuel, A., 1972. *Unequal Exchange,* (London: New Left Books).

Evans-Pritchard, E.E., 1973. Some reminiscences and reflections on fieldwork, *Journal of the Anthropological Society of Oxford,* 4, pp. 1-12.

Frank, A.G., 1979. *Dependent Accumulation and Underdevelopment,* (London: Macmillan).

Gans, H., 1967. *The Levittowners,* (London: Allen Lane).

Golding, P. and Middleton, S., 1982. *Images of Welfare: Press and Public Attitudes to Poverty,* (Oxford: Basil Blackwell & Martin Robertson).

Harris, C., 1974. *Hennage: A Social System in Miniature,* (New York: Hart-Reinhold).

Lee, D. and Newby, H., 1983. *The Problem of Sociology,* (London: Hutchinson).

Lipietz, A., 1977. *Le Capital et Son Espace,* (Paris: Maspero).

Lipietz, A., 1980. The structuration of space, the problem of land and spatial policy, in: J. Carney *et al.* (eds.) *Regions in Crisis,* (London: Croom Helm).

Lowe, P. and Goyder, J., 1983. *Environmental Groups in Politics,* (London: Allen & Unwin).

Mackay, G.A. and Laing, G., 1982. *Consumer Problems in Rural Areas,* (Edinburgh: Scottish Consumers Council).

Markusen, A.R., 1980. Regionalism and the capitalist state, in: P. Clavel *et al.* (ed.) *Urban and Regional Planning in an Age of Austerity,* (New York: Pergamon).

Massey, D., 1979. In what sense a regional problem?, *Regional Studies,* 13, pp. 233-243.

Massey, D., 1981. The UK electrical engineering and electronics industry: the implications of the crisis for the restructuring of capital and locational change, in: M. Dear and A. Scott (eds.) *Urbanization and Urban Planning in Capitalist Society,* (London: Methuen).

Massey, D., 1982. Industrial restructuring as class restructuring: productive decentralisation and local uniqueness, *Regional Studies,* 17, pp. 73-89.

Moseley, M., 1979. *Accessibility: The Rural Challenge,* (London: Methuen).

Moseley, M. and Packman, J., 1983. *Mobile Services in Rural Areas: Final Report to the Department of the Environment,* (Norwich: University of East Anglia).

Newby, H., 1977. *The Deferential Worker,* (London: Allen Lane).

Newby, H., 1981. Urbanism and rural class structure, in: M. Harloe (ed.) *New Perspectives in Urban Change and Conflict,* (London: Heinemann).

Newby, H., Bell, C., Rose, D. and Saunders, P., 1978. *Property, Paternalism and Power,* (London: Hutchinson).

22

O'Brien, P.J., 1975. A critique of Latin American theories of dependency, in: T.B. Oxaal and D. Booth (eds.) *Beyond the Sociology of Development,* (London: Routledge & Kegan Paul).

Pahl, R., 1965. *Urbs in Rure,* (London: Weidenfeld & Nicholson).

Pahl, R., 1966. The rural-urban continuum, *Sociologia Ruralis,* 6, pp. 299-327.

Pahl, R., 1967. The rural-urban continuum: a reply to Eugen Lupri, *Sociologia Ruralis,* 7, pp. 21-29.

Phillips, D. and Williams, A., 1982. *Rural Housing and the Public Sector,* (Aldershot: Gower).

Pickvance, C., 1983. *Spatial policy as territorial politics: the role of spatial coalitions in the articulation of 'spatial' interests and in the demand for spatial policy,* (paper given at British Sociological Association Annual Conference, Cardiff).

Powdermaker, H., 1966. *Stranger and Friend: The Way of an Anthropologist,* (New York: Norton).

Rapoport, R.N., Fogarty, M.P. and Rapoport, R., 1982. *Families in Britain,* (London: Routledge & Kegan Paul).

Rose, D., Saunders, P., Newby, H. and Bell, C., 1976. Ideologies of property: a case study, *Sociological Review,* 24, pp. 699-731.

Saunders, P., 1981. *Social Theory and the Urban Question,* (London: Hutchinson).

Sayer, A., 1979. Theory and empirical research in urban and regional political economy: a sympathetic critique, *Urban and Regional Studies Working Paper* 14, (Brighton: Sussex).

Shucksmith, M., 1981. *No Homes for Locals,* (Farnborough: Gower).

Smith, M.P., 1980. *The City and Social Theory,* (Oxford: Basil Blackwell).

Stacey, M., 1969. The myth of community studies, *British Journal of Sociology,* 20, pp. 34-47.

Stein, M., 1964. *The Eclipse of Community,* (New York: Harper & Row).

Valentine, C.A., 1968. *Culture and Poverty: Critique and Counter-Proposals,* (Chicago: University of Chicago Press).

Wallerstein, I., 1974. *The Modern World-System: Capitalist Agriculture and the Origins of the European World-Economy in the Sixteenth Century,* (New York: Academic Press).

Williams, R., 1973. *The Country and the City,* (London: Oxford University Press).

Winter, H., 1980. *Homes for Locals,* (Exeter: Community Council of Devon).

SECTION I

THE POLITICAL ECONOMY

OF

RURAL REGIONS

2. Rural regions in national and international economies

GARETH REES

Changes in rural employment structures are central to any understanding of the reality of rural social life. On the one hand, they reflect profound shifts in the nature and organisation of capitalist production and, more specifically, the widely differing impacts of these shifts on different types of locality. On the other, employment changes themselves have resulted in radical developments in terms of rural class structures, gender divisions, the forms of political conflict occurring in rural areas and, indeed, of the complex processes by which 'rural cultures' are produced and reproduced.

I shall be concerned to root my more general and abstract arguments empirically in the Welsh experience of rural change. What has happened in rural Wales illustrates with great clarity the broader issues with which I shall be principally concerned, as Tom Nairn (1977) has commented (in a rather different context): "In the Welsh knot the usual forces of uneven development have been tied together unusually closely and graphically" (p. 211).

What follows will be roughly divided into three parts. Firstly, I shall review briefly some of the theoretical background to what I have to say and, in particular, examine what Newby and Buttel (1980) have termed the 'new' or 'critical rural sociology'. In section two, I shall sketch out some of the major trends in the development of rural employment structures, relating these to wider changes in capitalist production and drawing heavily on the historical experience of rural Wales. In the third section, I shall refer very briefly to some of the implications of these changes for the social structures of rural localities more generally.

THEORETICAL CONSIDERATIONS

It is by now a common-place that much of what has passed for rural sociology is largely discredited. In a recent review essay, Bradley has summarised the situation succinctly:

27

Is there a rural sociology? In the dozen
years or so since Manuel Castells first
sounded the death knell for urban
sociology, scarcely any Marxist writing
has appeared outlining a political economy
or rural society under advanced capitalism.
During this same period rural sociologists
have experienced a profound sense of inner
disquiet. Conventional tools seem insufficient to answer basic questions. The
"rural-urban continuum" has generally been
discarded as geographical determinism;
diffusionist studies of farmers' responses
to new technology appear dated; and the
production of academic texts reifying the
"traditional" values embodied within
rural communities is *passe*.
(Bradley 1981 p. 581)

Equally, however, there are clear indications that new
perspectives are emerging, to a large extent in conscious
reaction to the sterility of much of this past work. Hence,
for example, Newby and Buttel (1980) have argued that,
under the influence of 'neo-Marxist and kindred perspectives',
a new research agenda for rural sociology has emerged which
encompasses: "the structure of agriculture in advanced
capitalism, state agricultural policy, agricultural labour,
regional inequality, and agricultural ecology". Moreover,
there is already a substantial body of research which
demonstrates the fruitfulness of such new approaches; for
example, the Essex studies of class relations within agricultural production (Newby 1977, Newby *et al*. 1978); analyses
of developments in capitalist agricultural production
(Friedland, Barton and Thomas 1981); and historical and
comparative studies of rural social change (Newby 1978,
Buttel and Newby 1980).

This transition has been acutely experienced in the
analysis of the Welsh social structure (Day 1979). In so
far as there has ever existed a distinctive, indigenous
tradition of social analysis of Wales, it has been dominated
by a perspective derived from studies of rural communities
(for example, Rees 1950; Davies and Rees 1960). These
studies delivered a view of the Welsh social structure
which focussed upon the enduring charcteristics of 'community':
internally coherent, stable and with few tensions and conflicts. Moreover, the mainspring of these characteristics
was the adherence to the Welsh 'way of life', rooted in
traditional rural society and centred upon the cultural
configurations of nonconformity and the chapel, the Welsh
language and literary tradition, traceable back over many
centuries.

Given this ahistorical and functionalist view, when
change did occur it was necessarily presented as something
which was imposed from outside and threatened the vitality
of 'real' Wales. In the upland, rural areas, these intrusions

were seen to arise from the Anglicisation of the gentry
and the alien values this introduced. More significant,
however, was the debilitating influence exerted by the growth
of industrial and urban Wales, largely in the south; a
growth which, though engaging the vast majority of the
Welsh population, could be defined as alien and un-Welsh –
a threat to the essence of Welsh social life.

What is involved here, of course, is a particular
expression of much wider debates. The picture of Wales
which emerges resonates clearly with the more general discussions of *Gemeinschaft-gesellschaft,* the urban-rural
continuum, the rural 'way of life' and so on. What is
distinctive, however, is that a version of rural society is
equated with the *reality* of the whole Welsh social formation.

Now, given the particular shape which this broader
debate has taken in Wales, it is not surprising that the
reaction against the older tradition has been especially
vehement (although, of course, this tradition continues to
exert a powerful influence within certain strands of political thinking in Wales) and that it has focussed upon the
development of an adequate analytical framework for the
sociology of Wales, rather than of rural society *sensu
strict.* Hence, intellectual effort has centred upon trying
to understand the Welsh situation as an instance of more
general patterns of 'regional' inequality, with relatively
little attention being paid to the specificities of rural
localities within Wales.

Initially, much of this intellectual effort involved
the attempt to use models formulated within the sociology
of development to understand the dynamics of neo-colonialism,
in order to analyse *regional* inequalities *within* the
advanced economies. Much of the initial impetus here derived
from Hechter's (1975) study of the historical relationships
between the English 'core' and the Celtic 'periphery' of
Ireland, Scotland and Wales in terms of 'internal colonialism'
and his notion of a 'cultural division of labour'. Latterly,
however, attention has shifted to the work of 'dependency'
theorists, initially formulated in the context of Latin
American development. At its most basic, the central claim
here is that – to use Frank's (1967) initial formulations –
the 'development' of the 'metropolis' necessarily implies
the 'underdevelopment' of the 'satellite'. Inherent in
capitalist accumulation is a chain of extraction of surplus
from the satellite to the metropolis; a chain which determines
both international relationships within the world economy
and also the internal structure of the satellite itself.
(Carter 1974, applies essentially this perspective to an
analysis of the Highlands in relation to the development
of the wider Scottish society.) However, as more recent
analysis has shown, the adoption of a dependency framework
does not necessarily imply a wholly static relationship of
exploitation between the dependent and the advanced countries
or regions. As both Lovering (1978a) and Cooke (1980)
have argued for the Welsh example, quite considerable

advanced economic development may occur within the dependent
economy, but such 'enclave' development takes place on terms
set by metropolitan capital and remains almost wholly
externally controlled. Again, this has clear implications
in terms of patterns of change internal to the dependent
economy, suggesting the emergence of an interdependent but
bifurcated social structure.

Time does not permit a detailed evaluation of these
important theoretical developments — all of which, it should
be noted, have potential applications to the analysis of
rural social change. However, it is important for my general
arguments to summarise some of the central difficulties and
problems. (For excellent extended discussions, see Massey
1978, Lovering 1978b, Day 1980.) Three issues are especially
significant. Firstly, such approaches run an acute risk of
fetishising spatial units. Urry (1981) has made the point
well:

> It is illegitimate to talk as though
> there were an interdependence between
> different spaces *per se*. Spatial patterns
> cannot be said to interact, only the
> social objects present within one or more
> such spaces. It may therefore be incorrect
> to talk of one area exploiting another area;
> to suggest that a region is exploiting
> another region, or that the centre is
> exploiting the periphery, may be to
> fetishize the spatial.
> (p. 457)

What this implies, of course, is that when such formulations
are used, they gloss over the real causal mechanisms which
in fact produce the observable social phenomena, encapsulated
in concepts such as regional inequality or rural-urban
contrasts. Thus, to define issues in these sorts of spatial
terms is in truth to mis-specify them; and this in turn
means diverting attention from the reality of the underlying
processes.

Secondly, and closely related, is the argument that
dependency approaches are in danger of displacing the study
of class relationships from the centre of the analysis.
As Brenner (1977) has remarked, the inadequacy of many
dependency approaches lies in their failure to specify:

> the particular, historically developed
> class structures through which these
> processes actually worked themselves out
> and through which their fundamental
> character was actually "determined".
> (p. 91)
> (See also Day 1980, p. 243)

It is, of course, true that the analysis of capitalism as a
mode of production does yield certain expectations as to

general lines of development. However, this is insufficient to understand actual historical outcomes, which are dependent also on particular configurations of economic, political, ideological and other conditions.

Thirdly, as Day (1980) has pointed out, there is a tendency within dependency models (as with more traditional analysis of regional inequality) to take as their *starting point* the recognition of the unevenness of development. Unevenness itself remains largely unexamined.

The *positive* implications of these criticisms can be seen in the emergence of analyses of regional inequality which may be loosely grouped under the label of 'uneven development'. The starting point here is the nature of capitalist production itself and, more particularly, what Marx identified as the central imperative to expand upon capital accumulation and to attempt to maximise the extraction of surplus value.

Capital accumulation involves the conversion of a portion of surplus value which the capitalist expropriates from the workers into additional capital, whether as labour or as fixed capital and raw materials. This enlarged capital is thus enabled to extract higher levels of surplus value and capitalists thus become engaged in a ceaseless process of re-investment. This process, in turn, is necessitated by the inherent competition between individual capitals and the necessity of achieving at least normal profits in order to survive. In this way, each individual capital will constantly seek out new opportunities for profit, thereby generating spontaneous, unregulated patterns of growth and contraction, which in their very nature are unbalanced.

Periodically, of course, crises will occur as barriers to further expansion of accumulation are reached such crises will find expression in falling profit rates, excess production or underinvestment. Hence, it will become necessary to restructure production to allow accumulation to proceed. Such restructuring may take a variety of forms: the raising of productivity through the introduction of new production techniques or of new methods of labour control; the reduction of labour costs; the extension of markets; or the destruction and replacement of existing capital. Crucially for present purposes, this restructuring may involve the *geographical* reshaping of productive activity, to take advantage of openings for greater profits.

Hence, during the course of development, the general conditions of production will change, dependent upon tech-nological innovation and alterations in the balance of class relations. However, the significant point is that at any particular moment, the conditions to which investment must respond will include the results of earlier historical changes, including the pre-existing spatial distribution of the means of production. As Massey (1978) argues:

the process of accumulation within
capitalism continually engenders the
desertion of some areas, and the creation
there of reserves of labour-power, the
opening up of other areas to new branches
of production, and the restructuring of
the territorial division of labour and
class relations overall.
(p. 106)

The end result, then, is a pattern of regional effects in
which:

the social and economic structure of any
given local area will be a complex result
of the combination of that area's succession
of roles within the series of wider national
and international spatial divisions of
labour.
(Massey 1978, p. 116)

Moreover, it should be emphasised that class relationships
are not a kind of optional extra to be added on to this
analysis as desired. Capital accumulation does not take
place in a vacuum. It takes place in localities with
differing histories, class compositions and social characters
generally. Such differences are, in one sense, the *products*
of the accumulation process. Equally, however, they *shape*
the direction and pattern of the accumulation process.
Hence, analysis in this mould concerns itself not simply
with changes in the nature of production, but also with the
class structure, political movements, the role and functions
of the state and a host of other topics. The essential
coherence of this work derives from its starting point in
accumulation.

Now, the point of spending a little time reviewing
these theoretical developments in the sociology of Wales
and regional development more generally is that it casts
new light on the 'critical rural sociology' which I referred
to earlier. Here the emerging emphasis appears to lie in
understanding rural society in terms of the analysis of
agricultural production and the social relations thereby
generated. The specificity of rural localities is seen to
derive from the peculiarities of the capitalist production
of agricultural commodities, with issues of landownership
consequently occupying a central analytical position.
Newby (1978) has made these points more or less explicit:

The "rural" need not, of course, be
regarded as coterminous with the
"agricultural" but all meaningful definitions of "rural" have at least a basis
in agriculture and so a consideration of
the development of agriculture remains
at the heart of any rural sociology. ...
Where rural sociology can establish itself

as a viable sub-discipline is by basing
its enquiries upon a recognition that
agriculture does *not* develop in advanced
capitalist societies in a manner which
merely mirrors other industries. Much
of the distinctiveness of agriculture ...
derives from the importance of land as a
factor of production. If for no other
reason, agriculture therefore throws
into sharp relief the nature of property
relationships in capitalist societies
and much rural sociology could be
profitably reoriented to a consideration
of property, which continues to remain an
underexamined aspect of modern societies
generally. In addition, the importance
of land as a factor of production increases
the spatial constraints upon the social
and economic organisation of agriculture.
(p. 25)

In the light of my earlier discussion of the literature
on regional development, these arguments can be restaged
so that the generation of 'rural localities' may be viewed
as the outcome of the use of space by particular fractions
of capital. The predominance of agricultural production in
given areas is the result of investment patterns reflecting
specific dimensions of the accumulation process and the
searching out of profits; as are changes in agricultural
technology and the organisation of farming more generally.
In other words, agricultural production and its associated
spatial manifestations are nothing more than a particular
instance of much more general trends within capitalist
production and the uneven spatial development thereby
engendered. Indeed, it is perhaps doubtful that agriculture
has even the degree of specificity which Newby claims for
it (i.e. in terms of the significance of land) when compared
with other branches of capitalist production such as coal-
mining, iron and steel production and so forth.

This sort of reformulation - which I take to be wholly
compatible with the 'new rural sociology' - is, I think,
entirely helpful in terms of the analysis of agricultural
production. However, there seem to me to be dangers in too
readily equating the analysis of 'rural localities' with the
analysis of agriculture. This point here can be made most
readily in terms of Massey's (1979) extremely useful image
of the determination of the character of particular localities
as the outcome of 'rounds of investment', reflecting the
area's role in a succession of different spatial divisions
of labour. Thus conceptualisations of the 'rural' in terms
of the dominance of *agricultural* production in particular
localities are in fact based upon the outcome of *past*
rounds of investment. There is no reason, of course, why
present or future rounds of investment should not engender
radically different spatial configurations of productive
activity, thereby producing radically different social

structures too. The problem, then, is that by focussing upon one particular aspect of capitalist production (agriculture) because this may have been the *historically* determining influence in given (rural) areas, attention may be diverted from the totality of contemporary processes operating in areas thus defined. Indeed, it may be that to begin from a conceptualisation in terms of rural localities may itself be misleading.

A POLITICAL ECONOMY OF RURAL WALES

I can perhaps illustrate the significance of these points by reference to the experience of processes of development in rural Wales. Historically, of course, a large part of Wales was dominated by agricultural production. And even after the penetration of industrial capitalism, mainly during the nineteenth century, creating in south Wales what Nairn (1977) has described as 'a great secondary centre of the European industrial revolution', mid, west and north Wales remained areas largely of farming. Indeed, it was this fact which provided the material basis for the kind of rural community studies which I described earlier.

Needless to say, of course, as Adamson (1981) has shown, the reality of social relationships even in nineteenth century rural Wales was far from the picture painted in such studies. Indeed, he has argued convincingly that the dynamic of change in agriculture during this period is best understood in terms of the conflicts between distinct class groupings, defined in terms of their relationships to the ownership and control of land; conflicts between, then, a land-owning gentry, small tenant farmers and agricultural labourers. Moreover, these conflicts themselves occurred in a context which was set by the wider imperatives of shifts in the profitability of particular forms of agricultural production and the organisation of land-holding.

More recently, too, the social structure of rural Wales has been greatly influenced by general trends in the structure and organisation of farming. In particular, three factors have been important. Firstly, there has been a break-up of the great rural estates and the transformation of patterns of landownership, with former tenants tending to buy their (relatively) small holdings. Hence, between 1909 and 1941, the proportion of Welsh agricultural holdings occupied by the owner increased from some 11 per cent to 37 per cent; by 1960, the proportion had reached 58 per cent; and by 1970, 64 per cent (·Williams 1980). Second, the size of agricultural units has increased: between 1951 and 1971, the number of units larger than 100 acres increased by 17 per cent; whilst those of less than 20 acres fell by over 70 per cent (Williams 1980). Thirdly, there has been a massive substitution of capital for labour in Welsh agriculture. During the fifty years after 1921, the numbers of agricultural workers in Wales fell from some 107 000 to less than 53 000 (in spite of a growing total workforce) (Williams 1980).

Though these developments are of the utmost significance
in determining the political economy of rural Wales, an
overly exclusive focus on changes within agriculture will
result in only a partial understanding of that political
economy and the processes by which particular localities
acquired their distinctive character. What is crucial here
is the interaction between the historically constituted
agricultural structure and developments in other sectors
of the economy.

I think that this is true even of much of nineteenth
century Wales. For example, in north-west Wales, during the
latter part of the eighteenth and early nineteenth century,
a close interdependency developed between agriculture and
what was then an extremely 'modern' industry - slate quarrying
and mining. The connection lay in the means by which the
quarry-owners recruited their labour and kept their wages
down to a minimum. Hence, over wide areas of the Vaynol
and Penrhyn estates, the custom was to provide new quarrymen
with plots of land to live on, sometimes with building
materials provided or with an existing building included.
During the earlier part of the century the quarryman would
generally produce his family's food and clothing from this
plot, with his earnings from the quarry being used mainly
to pay the rent and to buy extras from the towns. For the
quarryman, this system provided the essentials of survival;
but for the landlord-capitalist it created a source not
only of enormous profit, but also tight control over the
workers (Jones 1981; Lovering 1982).

Thus, for this particular part of early nineteenth
century rural Wales, it is not possible to comprehend the
nature of the local social structure without taking account
of the totality of roles which the area played within a
number of spatial divisions of labour, only one of which
was related to agriculture. Moreover, as Jones (1981)
shows, the complex interdependencies of these roles yielded
a 'culture' which far from being a quaint oddity in a
marginal backwater, actually generated political struggles
which occupied a significant position in the development
of the British labour movement. For example, during the
lockout from the Penrhyn quarries between 1900 and 1903,
£300 was raised in Bethnal Green in east London for the
struggle of the quarrymen and their families. Of this,
£25 was subscribed in farthings; 24000 people would have
to give a farthing to reach this figure!

Post-war developments in rural Wales illustrate the
general argument with even greater clarity. Quite simply,
profound shifts in the organisation of capitalist production
are radically reshaping historically-constituted rural
areas. Moreover, many of the principal changes taking
place are not within agriculture, but within other sectors
of the capitalist economy. One of these changes has been
quite widely recognised within the rural sociological
literature - the trend toward the location of large, highly
capital-intensive developments, such as nuclear powerstations,

water supply schemes, oil terminals and refineries, and smelters, in remote, sparsely populated areas (see, for example, Davies 1978; Markusen 1980).

Some progress has been made toward understanding the essential features of such investment. In particular, it is quite well established that the employment effects on the receiving localities are very small in the longer term. As an Economist Intelligence Unit report (1972) commented on the basis of an analysis of major construction schemes in north Wales:

> Numerically by far the larger part of the labour force is unskilled and semi-skilled. Skilled jobs are created, and there is a considerable supervisory, technical and management force. The projects are undertaken by large const-ruction companies - either individually or in consortia. These companies maintain permanent "senior" staff who move around the country to work for their companies in their current contracts, and are likely to make up the entire management force of the projects. Similarly teams of 'specialist' skills will be maintained, and to a lesser extent teams of skilled and semi-skilled labour will work with one company, again following it from project to project. The composition of these teams is likely to be fairly mobile, with labour recruited locally following the company to a new scheme elsewhere at the termination of the existing work; and similarly labour recruited elsewhere remaining in the area after work has finished.
> (p. 12)

Given this employment pattern, local labour is likely to be taken on only in the unskilled categories. Moreover, recruitment creates direct competition with pre-existing economic activities for labour resources. Hence, the Economist Intelligence Unit team found that agricultural workers were being drawn into construction schemes, resulting in the permanent loss of agricultural employment as farmers compensated by substituting capital for labour; even though, of course, jobs in the construction schemes were transitory. In the longer term, this led to movement out of the area, with many local workers moving on to the next construction site. (Markusen 1980, reports similar findings in her study of the 'boom towns' of the western USA.) In short, the employment effects are likely to be the destabilisation and disruption of pre-existing employment patterns. These effects are likely to impact not only upon local labour, but also upon local, small-scale capital, thereby providing some basis for the formation of political movements in opposition to developments of this kind.

Though a great deal more could be said about this
sort of development, I want to concentrate in the remainder
of this paper on much more pervasive and general trends
in manufacturing and services which are occurring in what
would be historically constituted as rural Wales. A
cursory examination of the aggregate employment statistics
for rural Wales reveals at least two major features.
Firstly, throughout the post-war period, by far the dominant
sources of employment have been in manufacturing and services.
Secondly, this dominance has increased through the period
not simply as a result of labour-shedding in the primary
sectors, but also - and more interestingly - because of the
really quite startling increases in employment which have
occurred particularly in manufacturing, but more latterly
in services too. Hence, for example, Fothergill and Gudgin
(1979) report percentage increases for manufacturing
employment in sub-regions of rural Wales between 1959 and
1975 ranging from 53 to 313 per cent (in north-west and
mid Wales respectively). Less dramatically, Department of
Employment figures reveal increases in service employment
in parts of rural Wales ranging up to 20 per cent for the
period 1971 to 1977.

Given the population bases of these areas, the absolute
numbers of workers involved in these changes are small,
compared with, say, the parallel changes in industrial
south Wales. Nevertheless, for the localities concerned
and their residents, they are indicative of the profoundest
of social changes, consequent upon the emergence of new
roles within particular spatial divisions of labour for
these areas. Moreover, the importance of these Welsh
changes is heightened by the fact that they are but particular manifestations of much broader shifts affecting the
disposition of economic activity over the capitalist
economies of Western Europe and North America, not just
Britain. Perhaps the more familiar aspect is the decline
of manufacturing in major urban areas. However, as
Fothergill and Gudgin (1982) write - in a chapter called
'Urban Decline and Rural Resurgence':

> The decline of cities and the growth of
> small towns and rural areas is the dominant
> aspect of change in the location of manuf-
> acturing industry in Britain and other
> Western industrial economies. The strength
> and pervasiveness of this urban-rural
> shift has been remarkable, and there is
> every prospect of it continuing. ... the
> contrast in manufacturing employment
> change between cities and small towns has
> been very large - much larger than the
> differences between regions - and very
> consistent. The larger and more industrial
> a settlement, the faster its decline. At
> the two extremes, London lost nearly 40 per
> cent of its manufacturing jobs between
> 1959 and 1975, while the most rural areas

increased theirs by nearly 80 per cent
during the same period.
(p. 68)

The generality of this pattern of development, however,
does not imply that there is any emerging homogeneity in
historically-constituted rural areas. Their pre-existing,
social structures are diverse and the spatial distribution
of manufacturing shift is itself uneven. Kennett and Hall
(1981) have described the areas of greatest growth in the
1970s as:

> a broad contiguous belt from the South
> Coast in east Dorset and Hampshire, taking
> in part of the South West region (Avon
> and Wiltshire), the western and northern
> sectors of the Outer Metropolitan Area
> and the Outer South East, with a spur
> running south of London to the Crawley-
> Burgess Hill area of Sussex, and thence to
> East Anglia and the northern part of the
> East Midlands region. ... But even outside
> this favoured belt, there were gains in
> employment in more distant rural areas
> ...: Northumberland, North Wales, and
> Highlands of Scotland.
> (p. 35)

Underneath this broad regional characterisation, however,
there was an enormous complexity of highly localised changes.
To return to the Welsh example, much of the manufacturing
development in mid Wales has taken place in Newtown, under
the auspices of the Mid Wales Development Corporation and
subsequently the Development Board for Rural Wales; with
workers being drawn in from surrounding villages over rather
a wide area (see, for example, Williams 1980). Similarly,
Lovering (1982) in a detailed analysis of changes in the
economic structure of Gwynedd, has shown how new manufacturing plants have tended to cluster to particular localities
within the region. The picture that emerges is of a patch-
work of interrelated local labour markets, each one evolving
a particular character as a result of much wider shifts
in capitalist production. This complexity of effects is
mirrored by a complexity of determinants. As Massey and
Meegan (1982) have argued, aggregate shifts in employment
structure mask a diversity of real processes operating
during different time periods, in different industries and,
indeed, in different firms within the same industry. It
would, of course, be quite impossible to encompass this
complexity in a short paper.

However, I want to conclude this discussion of manufacturing change by reference to *one* of the broad trends of
industrial reorganisation, which appears to have had a
particular significance for rural Wales. Urry (1981) has
identified the key significance in the determination of
local changes of:

the increased concentration and centralization of capital nationally and
internationally; this leads to (a) an
increase in the degree to which capital
can redistribute its activities in order
to take advantage of all possible variations
in the price, availability, skills and
organization of the local labour force
which remains relatively immobile and
(b) a decline in the interlinkages
between local/regional capitals and an
increase in the "external control" of each
local economy.
(p. 464)

Lovering (1982) describes the effects of precisely such
changes in the labour market of Gwynedd. A considerable
proportion of the inward investment which has taken place,
accounting for the overwhelming majority of manufacturing
employment growth in the area, reflects the growing inter-
nationalisation of the UK economy generally and the
increasing significance of multi-regional, British companies.
During the earlier post-war period, companies such as
Bernard Wardle, Ferrantis, Lairds, Hotpoint, Greengate and
Irwell and Ferodo all established operations in the area.
Latterly, these have been joined by Anglesey Aluminium
(part-owned by the Kaiser Corporation), J.P. Wood (Chukie
Chicken), Peboc, Hi-speed Plastics and Smith-Corona. So
striking were these developments that a research team from
University College, Bangor claimed of Anglesey's economy
that there was a growing polarisation between the 'old' and
'new' sectors:

the economy of the region will gradually
split and the new industry and its
attendant labour force will gradually
become an enclave in an economy to which
it has little economic relation.
(Sadler, Archer and Owen 1973, p. 78)

This polarisation is real enough. Quite apart from
corporate structures, there are contrasts in terms of
product types, methods of production and, in consequence,
types of employment. Conditions of work in the 'new'
sector certainly appear to be superior to those in the
smaller-scale, locally-controlled plants. However, this
should not disguise the poverty of employment in the 'new'
sector too. Again, Lovering's (1982) study reveals useful
evidence:

In keeping with the pattern of manufacturing
change in the UK generally many of the jobs
created in these new factories bore little
resemblance to the proud myths of modernis-
ation - instead of creating a workforce of
highly skilled and well-paid workers they
concentrated on routine unskilled or semi-
skilled work, often part-time and frequently

female because of the lower wages women
would accept. The Greengate and Irwell
Company, for example, commented that the
average working life of its "girls" was
3 years. This pattern militated against
trades unionism and many small firms
explicitly forbade union membership.
Beneath the illusion of industrial development, therefore, a new form of dependent
and segmented working class was being
created, while the prospects of self-
propelled future growth were minimal.
(p. 12)

The essential point is that within the wider context of the
re-organisation of capitalist production, certain types of
new jobs have been created in this, and other parts of rural
Wales (see, for example, Wenger 1980; Williams 1980); and
that these types of job have been created precisely because
of the pre-existing social structure of these areas, which
makes available a labour force extremely attractive to
capital.

These sorts of changes in rural Wales should not be
taken as representative of what is happening in all historically-constituted, rural localities. As a result of the
same developments in the general organisation of manufacturing production, areas in the southern part of England -
those which have in any case experienced the greater part
of employment growth - have also received a much wider
variety of jobs, with a greater representation of such
'higher order' functions as conceptualisation and design,
and management and control. Correspondingly, of course,
the introduction of such jobs is having effects on the pre-
existing social structures of these rural areas very
different from those in Wales. Indeed, such processes may
well constitute the material reality underpinning the
movement of 'city dwellers' into the countryside which has
been observed by numerous commentators (for example, Pahl
1965; Newby 1979).

I wish to conclude this section of the paper with some
very brief remarks about the second growth sector in the
employment structure of rural Wales: services. One dimension
of this growth has been in tourism, which, of course, has
been of particular significance in certain localities.
What is interesting here is that close parallels exist with
the pattern of development in manufacturing. There is, on
the one hand, an 'advanced' sector, reflecting major investments by large companies and consortia in marinas, hotels,
holiday camps and such like; on the other, there is the
small-scale, locally-controlled development - camp sites,
small shops, cafes, bed and breakfast, etc. However, in
both these sectors, the kinds of employment generated have
been for the most part of extremely low quality, whether
low-paid, part-time, female-dominated jobs or self-employment,
frequently undertaken in conjunction with other occupations.

However, over most of rural Wales, it is the state which has generated the majority of service jobs, in the various spheres of public administration and the welfare state, with complex effects. Given that such activity is not distributed on a strictly market basis, one effect has been to smooth out spatial inequalities (an effect comparable to transfer payments made by the state to welfare recipients). However, once again, the *majority* of jobs created in the states employ have been low-paid, routine and largely taken by women, although a small proportion of professional and middle-management jobs have also been produced – providing, incidentally, the only source of 'good quality' employment in rural areas for women, and in rural Wales, for Welsh-speakers. Nevertheless, some of the most disadvantaged workers in rural Wales are in state employment. This seems to me to be a problem of equal significance to that of the effectiveness of service delivery, which, amongst causes of rural deprivation, has preoccupied rural analysts hitherto.

CONCLUSIONS

In this paper, I have sought to illustrate the complexities of the changes taking place in rural Wales and rural local-ities more generally – changes which cannot be understood without reference to fundamental shifts in the nature and organisation of the totality of capitalist production. More specifically, I have argued that the pre-existing social structures of rural areas have provided conditions which have determined and shaped the pattern of change in that capitalist production. In doing so, these social structures have themselves been recast in ways that may make them less 'rural'.

What I have said, of course, has been confined to a time period which stops short of the most immediate past and, accordingly, has taken little account of the effects of the severe deepening of the British economic crisis since the late 1970s. It seems most unlikely, for example, that the patterns of employment growth in rural areas which I have described have survived this deepening crisis. Certainly, the statistics of unemployment in the rural parts of Wales would seem to bear this out: although it should be remembered that high levels of – particularly male – unemployment are entirely compatible with the trends with which I have been concerned. It remains to be seen what effects any easing of the crisis will have.

It is also true that I have not attempted to draw out systematically the effects which these changes have had on the social structures of rural areas. However, those effects are clearly there. Whole new class groupings are being introduced into rural areas; the traditional roles of men and women are being broken down; whole groups of workers are becoming marginalised from the economic mainstream in unemployment and are evolving new methods of 'making out' in the informal economy. All these changes, in turn, are being reflected in new cultural forms and new types of political conflict.

41

REFERENCES

Adamson, D., 1981. *Social class, religion and nationalism in the rural sector,* (Cardiff, Department of Sociology, University College, Cardiff, mimeo).

Bradley, T., 1981. Capitalism and countryside: rural sociology as political economy, *International Journal of Urban and Regional Research,* 5, pp. 581-587.

Brenner, R., 1977. The origins of capitalist development: a critique of neo-Smithian Marxism, *New Left Review,* 104, pp. 25-92.

Buttel, F. and Newby, H. (eds.) 1980. *The Rural Sociology of the Advanced Societies: Critical Perspectives,* (London: Croom Helm, and Montclair: Allanheld, Osmun).

Carter, I., 1974. The Highlands of Scotland as an under-developed Region, in: E. de Kadt and G. Williams (eds.) *Sociology and Development,* (London: Tavistock) pp. 279-311.

Cooke, P., 1980. Dependent development in United Kingdom regions with particular reference to Wales, *Progress in Planning,* 15, pp. 1-62.

Davies, T., 1978. Capital, state and sparse populations: the context for further research, in: H. Newby (ed.) *International Perspectives in Rural Sociology,* (Chichester: John Wiley), pp. 87-106.

Davies, E. and Rees, A. (eds.), 1960. *Welsh Rural Communities,* (Cardiff: University of Wales Press).

Day, G., 1979. The sociology of Wales: issues and prospects, *Sociological Review,* 27, pp. 447-474.

Day, G., 1980. Wales, the regional problem and development, in: G. Rees and T. Rees (eds.) *Poverty and Social Inequality in Wales,* (London: Croom Helm), pp. 230-251.

Economist Intelligence Unit, 1972. *Employment Consequences of Major Construction Works in North Wales,* (London: T.I.U. Ltd.).

Fothergill, S. and Gudgin, G., 1979. Regional employment change: a sub-regional explanation, *Progress in Planning,* 12, pp. 155-219.

Fothergill, S. and Gudgin, G., 1982. *Unequal Growth,* (London: Heinemann).

Frank, A., 1967. *Capitalism and Underdevelopment in Latin America,* (New York: Monthly Review Press).

Friedland, W., Barton, A. and Thomas, R., 1981. *Manufacturing Green Gold: capital, labor and technology in the lettuce industry,* (Cambridge: Cambridge University Press).

Hechter, M., 1975. *Internal Colonialism: the Celtic Fringe in British national development,* (London: Routledge & Kegan Paul).

Jones, M., 1981. *The North Wales Quarrymen 1874-1922,* (Cardiff: University of Wales Press).

Kennett, S. and Hall, P., 1981. The inner city in geographical perspective, in: P. Hall (ed.) *The Inner City in Context: The Final Report of the Social Science Research Council Inner Cities Working Party,* (London: Heinemann).

Lovering, J., 1978a. Dependence and the Welsh economy, *Economic Research Papers Reg 22,* (University College of North Wales, Bangor).

Lovering, J., 1978b. The theory of the internal colony and the political economy of Wales, *Review of Radical Political Economics,* 10, pp. 55-67.

Lovering, J., 1982. *Gwynedd in British Capitalism,* (paper presented to the Plaid Cymru Summer School, Trinity College, Carmarthen, mimeo).

Markusen, A., 1980. The Political Economy of Rural Development: the case of the Western U.S. Boomtowns, in: F. Buttel and H. Newby (eds.) *The Rural Sociology of the Advanced Societies: Critical Perspectives,* (London: Croom Helm), pp. 405-432.

Massey, D., 1978. Regionalism: some current issues, *Capital and Class,* 6, pp. 106-125.

Massey, D., 1979. In what sense a regional problem?, *Regional Studies,* 13, pp. 233-244.

Massey, D. and Meegan, R., 1982. *The Anatomy of Job Loss,* (London: Methuen).

Nairn, T., 1977. *The Break-Up of Britain,* (London: New Left Books).

Newby, H., 1977. *The Deferential Worker: A Study of Farm Workers in East Anglia,* (London: Allen Lane).

Newby, H. (ed.), 1978. *International Perspectives in Rural Sociology,* (Chichester: John Wiley).

Newby, H., 1979. Urbanization and the rural class structure, *British Journal of Sociology,* 30, pp. 475-499.

Newby, H., Bell, C., Rose, D. and Saunders, P., 1978. *Property, Paternalism and Power: Class and Control in Rural England,* (London: Hutchinson).

Newby, H. and Buttel, F., 1980. Toward a critical rural sociology, in: F. Buttel and H. Newby (eds.) *The Rural Sociology of the Advanced Societies: Critical Perspectives,* (London: Croom Helm), pp. 1-38.

Pahl, R., 1965. *Urbs in Rure* (London: London School of Economics, Geographical Papers, no. 2).

Rees, A., 1950. *Life in a Welsh Countryside,* (Cardiff: University of Wales Press).

Rees, G. and Rees, T. (eds.) 1980. *Poverty and Social Inequality in Wales,* (London: Croom Helm).

Sadler, P., Archer, B. and Owen, C., 1973, *Regional Income Multipliers*, (Bangor: University College of North Wales).

Wenger, C., 1980. *Mid Wales: Deprivation or Development,* (Cardiff: University of Wales Press).

Williams, G., 1980. Industrialisation, inequality and deprivation in rural Wales, in: G. Rees and T. Rees (eds.) *Poverty and Social Inequality in Wales,* (London: Croom Helm), pp. 168-185.

Urry, J., 1981. Localities, regions and social class, *International Journal of Urban and Regional Research,* 5, pp. 455-474.

3. Capitalist restructuring, recomposition and the regions

JOHN URRY

In this chapter I shall reconsider some of the recent literature on rural social relations in the light of contemporary debates on the process of capitalist restructuring. In particular, it will be argued that:

1. The most important developments within contemporary capitalism are not those which generate ever-increasing concentrations of capital, state power and labour power within urban rather than rural areas, and in which social relations in the latter are increasingly dominated by social relations in the former.

2. Nevertheless there are highly significant changes occurring within the time-space structuring of contemporary capitalist societies but their effects cannot be summarised in terms either of the dichotomy between rural and urban areas, or of apparently identifiable regions.

3. The most important of these changes involve the *spatial* restructuring of capitalist production *and* of civil society[1], and these patterns of spatial restructuring have had the effect of heightening the socio-political salience of local systems of social stratification.

4. In rural areas these local systems cannot be simply characterised in terms of the relations of a new 'middle class' to the existing rural class structure.

5. The stratification structure of any locality (whether formally urban or rural) is the interdependent effect of mutually modifying forms of structural determination, especially of the complex overlap between diverse spatial divisions of labour.

6. A most important process of contemporary change is what one might loosely describe as the ruralization of industrial/ urban relations. This process reinforces others which serve to fragment and decompose the industrial classes of contemporary capitalism, and to usher in a progressively de-industrialised economy and attendant social relations.

I will begin by identifying a number of difficulties within current rural sociology and shall then consider one important analysis of recent economic change which makes much of urban/rural differences. I shall then analyse changing patterns by which capital and labour-power are currently being reorganised. Finally, I shall consider the consequences for the generation of diverse local systems of social stratification.

PROBLEMS IN 'RURAL SOCIOLOGY'

In recent years rural sociology seems rather to have lost its way in contrast with urban sociology which has been transformed both by the neo-Marxist debates introduced by Castells, Harvey, Lojkine and Lefebvre, and more recently in the UK by the post-Marxist writings of Dunleavy (1980) and Saunders (1982). In effect, British rural sociology involves the following claims, which constitute its main features:

1. Property rather than occupation is the defining principle of rural societies and, hence, it is the organization of property relationships, rather than the division of labour, which shapes the rural class structure (Banaji 1976; Newby 1978, pp. 6-15; Stinchcombe 1961).

2. In British agriculture there has been a substantial shift away from the landlord-tenant property system towards that of owner occupation, as well as a large increase in the ownership of land by finance-capital (Newby 1978, pp. 12-15; Newby *et al*. 1978).

3. There is a profound and irreversible rationalisation of the agricultural industry, away from farming as a 'way of life' towards its organisation 'as a business'. In particular, there is the growth of so-called agri-business, although this is not to argue that small farms will necessarily disappear (Newby 1978, pp. 19-20; Newby 1979; Gasson 1966; Winter, Chapter 6).

4. In contrast with industry, the growth of mechanization has reduced the division of labour amongst agricultural workers. This is because there has been a fairly massive reduction in the labour force employed on each farm although little reduction in the tasks to be performed which are sequentially rather than concurrently organised (see Newby 1977 ch. 5; and Gasson 1980, on the effects of the sexual division of labour in farming.

5. This outflow of labour from agricultural employment has both undermined whatever solidarities previously existed within rural areas when labour was far more plentiful (Newby 1977, ch 5; Newby 1978, p. 21) and *reduced* bureaucratization on farms and, hence, the distance between farmers and their workers (Newby 1977, ch. 6; 1978, pp. 21-22).

46

6. There has been a marked expansion of an ex-urban middle class within rural areas and this has produced an 'encapsulated' rural community, particularly focussed on the 'farm' and defined partly by opposition to the 'newcomers' (Newby 1977; 1980b; Pahl 1965).

7. A major determinant of social relations within the countryside since the Second World War has been the state. The policy of agricultural protection has particularly benefitted large-scale capitalist farmers and landlords (Josling 1974; Newby 1980a, pp. 54-66), and that of countryside protection has restricted economic growth and the development of competition for local labour (Newby 1980a, pp. 267). Moreover, urban newcomers have reinforced this by seeking to preserve, in Pahl's felicitous phrase, their village in the mind' (1965).

8. The analysis of rural social relations will only be successful if a more 'holistic' approach is adopted, such as that of centre, semi-periphery, and periphery, or that of 'internal colonialism' (Newby 1980a, section 5; 1982, pp. 157-159).

Clearly, some very important insights have been developed especially with regard to the social organisation of agriculture which has certain features derived from land as a distinctive means of production (see the discussion by Winter, Chapter 6). However, Newby, amongst others, is well aware of some significant deficiencies and he commends the development of a more holistic approach on two grounds: first, that this would render problematic the categories 'rural' and 'urban' by identifying social processes *common* to both; and second, that this would relate explicitly the social structure to the "spatial structure of regional development and underdevelopment" (1980a, pp. 92). There are, however, serious difficulties associated with this programme, particularly if it is being claimed that new 'holistic', 'regional', 'theoretical' analyses could be unproblematically added to the existing research. Partly this is because notions of 'centre-periphery' and 'internal colonialism' have themselves been severely criticised for their relatively ahistorical, static and functionalist character (see Cooke 1983).

There are also problems highlighted by the recent elaboration of a realist philosophy of science which suggests that we should make very clear the distinction between the causal powers of designated entities and the actual empirical events to which these entities contingently give rise (Bhaskar 1979; Keat and Urry 1982, postscript; Sayer 1982). Empirical events are the product of the complex interrelations between those entities whose causal powers are in part being realised (Urry 1984). Within this account the category of the 'rural' seems to constitute neither an entity with specifiable causal powers, nor a range of empirical phenomena which stands in a coherent relationship to particular causally powerful social entities.

It should be regarded, therefore, as a 'chaotic conception'. Inadequate theory will result if we try to generalise from such chaotically produced empirical phenomena. Rather it is necessary to abstract from empirical phenomena in order to arrive at theoretically informed analyses of the causal powers of social entities, powers which only contingently generate empirical events. A non-chaotic conception of the 'rural' could be based on one of the following:

1. The 'rural' refers to all those areas in which agricultural production dominates the local economy, either because there is no manufacturing and service production, or because any that is present is dominated by the social relations of agricultural production, relations which stem in particular from land as the centrally significant means of production.

2. The 'rural' refers to a particular structuring of local civil societies in which the patterns of social reproduction and social struggle are structured by the class relations engendered by ownership and control of the agricultural means of production.

3. The 'rural' refers to those areas in which the density of the population is so low (whether or not because agriculture is the predominant industry) that the means of 'collective consumption' cannot be provided economically within that area and have rather to be located in non-rural, urban areas instead.

Finally, it should be noted that the legacy of the rural-urban continuum can be seen in the tendency to analyse the degree to which agricultural production is like, or is becoming like, industrial production (see Newby 1978, p. 25). Such an approach has two deficiencies: first, since industrial or manufacturing production only accounts for a quarter of the presently employed population in the UK, it is more important to consider any similarities with service production and employment; and second, since a growing proportion of labour is being carried out either within households (the self-service economy - see Gershuny 1978), or outside the formal economy, or within part-time employment (Pahl 1980; Urry 1983a), then it is agricultural production, especially in its simple commodity form, towards which at least some forms of urban-based labour are moving. When Kautsky talks of "a suppression of the separation of industry and agriculture" (Banaji 1976, p. 47) it may have less to do with agricultural production becoming like industrial production and more to do with major changes in the entire organisation of work in a de-industrialising society (see Urry 1983a).

British rural sociology, though, has failed to examine the changing economic and spatial structuring of manufacturing and service industry. This failure is especially striking since, in a recent analysis of such changes, Fothergill and Gudgin (1982) have employed the distinction between urban and rural areas as a major explanation of the spatial restructuring of industry. They maintain that recent patterns of employment change cannot be analysed simply in terms of north versus south, or in terms of distinctive regions. They say: "industrial structure has become more or less irrelevant as an explanation of disparities in regional growth" (1982, p. 59). Instead, they note that all the areas that experienced major employment loss between 1959 and 1975 contain a major conurbation, whereas many of the areas that gained employment in the same period are 'rural' (1982, p. 14). This contrast is particularly marked for manufacturing employment.

Moreover, if we break down the regions into various sub-regions then we find that there is a general relationship between the size of settlement and manufacturing growth - small cities grew faster than large cities; small towns grew faster than larger towns. Fothergill and Gudgin conclude that "the shift from urban to rural areas is the major trend in industrial location in Britain" (1982, p. 24). The larger the settlement size the faster the decline in employment, especially of manufacturing. This is because the larger the city, the higher the rates of plant closure, the greater the losses through plant contraction, and the lower the rates of expansion of surviving firms (1982, p. 81). They maintain, somewhat implausibly that the shift of manufacturing employment out of large cities is mainly because of the great difficulty that firms in the larger urban areas have in physically expanding their plant compared with those in smaller settlements and in rural areas. The relative lack of physical limitations on spatial expansion in the less urban and the more rural locations is seen by Fothergill and Gudgin as the crucial factor in explaining variations in employment change.

In order to provide the rudiments of what I consider to be a more precise and satisfactory explanation of changes to the civil society of 'rural' areas I shall, in the next section, use Fothergill and Gudgin's analysis as an example of the misapplication of conceptions of 'urban' and 'rural' space.

CAPITAL, LABOUR POWER AND THE 'RURAL'

There are two particular difficulties in Fothergill and Gudgin's analysis. First, as we have seen, identifying a locality in terms of its rural/urban characteristics is far too simplistic. Second, they presume that the way to analyse industrial change is through identifying certain general processes which are then, to varying degrees, developed within any particular local economy (Murgatroyd and Urry 1983; Sayer 1982). Neither of these positions can be justified. In particular, any local economy should be viewed as the particular product of the overlap, in time and space, of the forms of capitalist and state restructuring within the pertinent sectors of extractive, manufacturing and service industry. As Massey argues: "the social and economic structure of any given local area will be a complex result of the combination of that area's succession of roles within the wider, national and international, spatial divisions of labour" (1978, p. 116). Thus, relevant analysis does not consist of identifying certain general tendencies which are more or less developed in different localities, depending upon whether that locality happens to be more rural or more urban.

Broadly speaking the 'restructuring' analysis, which I am arguing for here (cf. Storper 1981) involves the following claims:

1. There are a number of different patterns of economic restructuring, of different spatial divisions of labour.

2. These restructurings stem from changing patterns of capitalist accumulation and especially from the internation-alisation and fractionalisation of capital.

3. In particular, changes in economic location cannot be explained in terms of 'economic' or 'political' factors, but, rather, in terms of the complex forms of restructuring necessary for sustained accumulation.

4. These restructurings both result from changes in class struggles and, in turn, transform the conditions under which social relations within particular areas are reproduced.

5. Any area can only be understood as the product of its location within a number of overlapping spatial divisions of labour.

6. The resulting patterns of uneven development cannot be analysed simply in terms of regions and regional decline.

What, then, are the main forms of the spatial division of labour which may characterise any sector? The following are six important forms (derived from Massey 1981; Massey and Meegan 1982; Walker and Storper 1981; see Urry 1984 for further details):

(i) regional specialization - until the inter-war period many sectors were characterised by a high degree of specialization within particular regions (for example, cotton textiles and textile machinery within Lancashire, mining and shipbuilding within the North-East, arable farming in East Anglia);

(ii) regional dispersal - other sectors are characterised by a high degree of dispersal, including most consumer services, some producer services, certain manufacturing industries (such as food processing and shoe production) and mixed farming. Labour reductions in this case will take the form of intensification - that is relatively uniform cutbacks spread throughout the different regions;

50

(iii) <u>Functional separation</u> between management and research
and development in the 'centre', skilled labour in old
manufacturing centres, and unskilled labour in the 'periphery';

(iv) <u>functional separation</u> between management and research
and development in the 'centre', and semi- and unskilled
labour in the 'periphery';

(v) <u>functional separation</u> between management and research
and development and skilled labour in a 'central' economy,
and unskilled labour in a peripheral economy;

(vi) <u>division</u> between one or more areas, which are
characterised by investment, technical change and expansion,
and other areas where unchanged and progressively less
competitive production continue with resulting job loss.
The former may involve the development of new products as
well as new means of producing existing products.

   As we noted above we should not analyse a given area
as purely the product of a single form of the spatial
division of labour. To do so is, as Sayer points out, to
"collapse all the historical results of several intersecting
'spatial divisions of labour' into a rather misleading term
which suggests some simple unitary empirical trend" (1982,
p. 80). Rather any such area is, economically and socially,
the overlapping and interdependent product of a number of
these spatial divisions of labour and attendant forms of
restructuring.

   An important consequence of these processes is that
uneven development does not simply take the form of
regional inequality. This can be seen, firstly, by noting
the following observations about the North-West which,
according to Fothergill and Gudgin, was one of only two UK
regions to possess a 'regional' industrial structure. Even
here, though, there were the following assorted variations
in a number of indicators of economic structure: in the
percentage change in male employment between 1960 and 1977,
from minus 27.7 per cent (Liverpool) to plus 15.6 per cent
(Crewe); in female employment, from minus 33.5 per cent
(Rossendale) to plus 58.7 per cent (Northwich); and in the
1980 ratio of female to male employees, from 0.534
(Warrington) to 1.165 (Southport). Indeed, from their own
study Fothergill and Gudgin conclude that, with the decline
in distinctively regional patterns of inequality, there
are enormous, significantly local variations, and "much
greater contrasts within any region than between the regions
themselves" (1979, p. 157).

   The importance of these intra-regional variations is
also supported by the analysis of recent migration patterns
where it was found that "intra-regional shifts of population
have been shown to overwhelm inter-regional contrasts"
(Kennett 1982, p. 40). Such intra-regional variations,
moreover, have stemmed from the trend towards decentralisation
within, and deconcentration between, urban labour markets.

Kennett suggests that the longstanding drift to the south from peripheral regions is now less important than the centrifugal movement from cities which has spilled across arbitrary, regional boundaries (1982, p. 41). Considering just those local authorities enjoying Special Development Area status in 1982, highly diverse population shifts were experienced between 1971 and 1981, ranging from population losses of 10 to 16 percent (Knowsley, Liverpool) to population gains of 10 to 12 per cent (Kerrier, Anglesey). Hence, as Kennett says: "to make any meaningful interpretation of labour migration, local labour markets should be used" as the relevant unit of analysis (1982, p. 41).

Before elaborating some further reasons for the importance of local labour markets, three other points about recent population movements should be noted. First, there is an extraordinarily high rate of residence change; about five million people in Britain change where they live each year and this has obvious implications for class composition and recomposition (Kennett 1982, p. 47). Second, the nature of such composition and recomposition is also related to the patterns of migration flow into and out of any particular local economy and it is wholly inappropriate simply to consider *net* migration. Interestingly, contrary to neo-classical migration models, which postulate rapid in-migration in the prosperous labour markets and rapid out-migration in the least prosperous areas, there is in fact a strong positive relationship between in-migration and out-migration in different areas (see the scattergram in Kennett 1982, p. 42). Third, the general shift of population from 'urban' areas and especially from the conurbations to less 'urban', more 'rural' areas has not simply resulted from changes in relative labour demand due to economic restructuring. It has also stemmed from an increased privatisation of civil society - of a rejection of certain, urban-based socialised forms of reproducing labour-power - tendencies made possible by the widespread growth of private transport.

Thus, Fothergill and Gudgin's research on shifts in manufacturing and service industry, alongside other studies on the restructuring of regional and local economies, and analyses of recent trends in population growth and migration, all suggest that sub-regional *local* economies are of particular significance within the contemporary British economy. In a recent paper I argued that these forms of restructuring were producing new and significant local variations in class structures, an increasing significance of spatial deprivations based on the 'inferiority' of one's own class structure *vis-a-vis* other class structures, and an increasing importance of struggles centred around defending or recapitalising the locality *vis-a-vis* other local/regional/international structures (see Urry 1981b; Harris 1983; Urry 1983b; on the second point, see Donnison and Soto 1980).

The following points summarise the reasons why patterns of spatial unevenness should not be viewed as taking a regional form:

1. The concept of 'region' is conceptually arbitrary and problematic (Urry 1981b; Grigg 1969).

2. Pre-existing patterns of regional specialisation have become overlaid by new forms of the spatial division of labour (Walker and Storper 1981; Massey and Meegan 1982).

3. The development of national and international branch circuits of capital led to a marked decrease in the degree to which productive systems are focussed upon a particular region (Lipietz 1980).

4. The 'periphery-centre' pattern of new employment in the period from the mid 1960s to the 1970s produced a considerable reduction in regionally based variations of unemployment and economic activity rates (Keeble 1976, pp. 71-85; Dunford, Geddes and Perrons 1980, pp. 12-13; on recent 'regional' changes see Regional Trends 1982).

5. The major divisions in contemporary England no longer appear to be regionally based but are rather based on a three-fold division: between the South-East; what Donnison and Soto call 'middle England', i.e. Luton, Swindon, Coventry, Peterborough, etc.; and the old industrial north (1980, pp. 140-142).

6. Internationalised capital is now so constituted that it is both relatively spatially-indifferent as to location, and can distribute different parts of its global operations into different labour markets, so taking advantage of variations in the price, availability, skills and organisation of the local labour force. There is no reason why it will be regionally distributed (Westaway 1974; Massey 1981; Walker and Storper 1981; Urry 1981b).

7. There is increasing politicisation of economic change, that is, the allocation of economic activity (whether public or private) is significantly a matter of political organisation, although there are, as yet, no effective regionally-based organisations in the UK (Carney, Hudson and Lewis 1980).

Walker and Storper have neatly summarised the significance of some of these points within the USA:

> the past concentration of industry has created areas with the most experienced, skilled, well organised, high cost, and militant labour force; as a result many industries, not only those which are labour intensive, have found it advantageous to seek out greener pastures in the suburbs, small towns, the south and beyond. (Walker and Storper 1981, p. 496)

So far, however, I have considered these changes from the viewpoint of capital and the effects which it must necessarily bring about - and indeed this is something of a deficiency of much of the 'restructuring' literature. Nevertheless, it is crucially important to consider as well some aspects and effects which follow from the processes of production of wage-labour. The most important aspect of this is that, unlike other commodities, labour-power is not itself produced under capitalist relations of production (Lebowitz 1980; Urry 1981a). It is of course produced, but partly within domestic relations (within 'civil society') rather than within capitalist relations. The process of production involves not simply consuming commodities produced within the sphere of capitalist production, but rather through human labour systematically converting the use-values available for consumption into refreshed and energetic labour-power. Three aspects of this process are particularly noteworthy.

First, the fact that other inputs into the production of commodities, apart from land and unprocessed raw materials, are themselves capitalistically produced means that they are subject to a process of geographical levelling or homogenization. This occurs as the spheres of production and circulation are developed and generalised, first within national economies, and then across national boundaries. This means, both, that industrial plants have greatly heightened locational freedom and are much less tied to particular spaces; and that competitive advantage in location can primarily be gained by exploiting differences in labour supply. The latter includes, not simply the quantity and costs of labour-power within a given labour market, as neo-classical theory would propose, but also its skill level, the conditions under which its reproduction is effected, and its reliability and susceptibility to control (Walker and Storper 1981, pp. 497-500).

Second, the organisation of civil society is not something which simply mirrors the wider capitalist economy, as Aglietta seems to suggest in his otherwise provocative notion of a 'mode of consumption' (1979). Various forms of social struggle and practice should be viewed in part as attempts to maximise the distance between such an economy and civil society. For example Humphries argues that the nineteenth century working class family can be so viewed, as providing insulation from the anarchy and exploitative relations of the dramatically expanding capitalist economy (1977; and Urry 1984, on the dimensions involved in the spatial structuring of civil society).

Third, recent changes in the organisation of capital and of the state have changed the parameters within which such relations can be established and sustained. In particular, the growth of multi-plant enterprises and of national and international circuits of capital (rather than local/regional circuits) have reduced each individual centre of population to the status of a *labour pool*. The important

linkages within a town or city are those which pass through
the household, through civil society, and not through the
private or public enterprises located within that area.
The other linkages, involving the sale and purchase of
commodities between enterprises, occur *across* the urban
boundary. Cities are, thus, increasingly significant sites
for the production of wage-labour. They are sites within
which pools of labour-power are systematically created and
reproduced. The urban area is a system of *production,* a
relatively closed system comprised of a large number of
interdependent, relatively privatised households wherein
wage-labour is produced under conditions of systematically
structured gender inequality (Broadbent 1977). Cities are
not so much an interlocking economy of producing and consuming enterprises but a *community of subjects* who produce
and who consume in order to produce. Moreover, this
production is necessarily local, it is principally produced
for the *local* market and, as such, subject to the constraints
of time imposed by the particular relations between households and workplaces. Cities are viewed as relatively
independent labour pools, each comprised of a large number
of separately producing households, linked with each other
competing for urban space. A substantial shift in the
structuring of each urban locality has, therefore, taken
place. Previously such localities were integrated within
the production and reproduction of capital. However, as
each urban locality is reduced to the status of a labour
pool it ceases to be integrated within the production
process of capital but, instead, becomes the sphere for the
production of wage-labour, within the civil society.

These points help to explain why it is that 'rural'
areas have become important locations for capital investment
in recent years. International capital has been transformed,
first, through an increasing spatial indifference, and
second, by the fractionalising of its different global
operations. Potential plants are often relatively small
(even if part of massive multinationals) and capital will
be relatively indiffenent as to where they are located.
Hence, labour-power assumes a particular importance as to
location - and this includes differences in cost, skill,
control, and reproduction. Provided there is or could be
sufficient labour in a 'rural' area then expansion may well
take place in that (green field) site rather than in
alternative urban areas. Cities have become relatively less
distinctive entities, by-passed by various circuits of
capital and of labour-power. Civil society is thus extended and, as a result of private transport, typical spatial
constraints upon local civil societies are transcended.
Individual subjects can increasingly choose where their
labour-power is to be reproduced, in cities, or towns or
'rural' areas; and yet, at the same time, the organisation
of the resulting local civil societies assumes a particular
importance in the response of individual localities to
economic restructuring and change.

LOCAL STRATIFICATION STRUCTURES

Elsewhere I have proposed four distinct local class structures that could be found in urban areas in the UK (ignoring
ethnic differences) (Urry 1981b):

(i)   large national or multinationals as dominant employers -
smallish intermediate classes; large working class, either
male or female, depending on supposed skill level;

(ii)   state as dominant employer - largish intermediate
classes; declining working class; high employment of women;

(iii)   traditional small capitals as dominant employers -
large petty bourgeois sector; largish male working class;
lowish female employment;

(iv)   private service-sector capitals as dominant employers -
largish intermediate classes with high female component;
smallish working class.

Certain points of clarification should be added. First,
local *social* structures should be analysed as local civil
societies and not merely as local *class* structures. A
crucial, yet relatively unexplored, determinant of the
consequences of such structures is that of the recruitment
into, and expulsion from, distinct places within the social
division of labour. These processes of the formation and
reformation of social groupings involves analysing the
changing structure of *local* markets, one important feature
of which is that of geographical mobility within and between
such markets. The social structures of rural areas will be
exceptionally diverse because of both the variety of ways
any such area may be located within agricultural divisions
of labour, and because of the complex patterns of inter-
relationship between such an agricultural spatial division
of labour, and that area's location within other spatial
divisions of labour. Finally, the competition between
localities seeking, in Massey's phrase, to be 'struck by
the lightning' of outside capital becomes an important
focus of socio-political organisation within any locality,
as well as ensuring that some such localities become constituted as a spatial reserve army through, what Walker terms,
the 'lumpen-geography of capital' (1978, p. 32).

These points raise important issues related to the
changing patterns of labour market segmentation. Kreckel
(1980) has usefully distinguished a number of different
processes of segmentation: namely, demarcation (craft
unionism versus all other workers); exclusion (regularly-
employed adult white males versus those not so 'blessed');
solidarism (of workers employed in an enterprise, occupation
or industry); inclusion (protecting and encircling a skilled
sub-market through corporation or occupation-specific
qualifications);   and exposure (of non-organised groups to
easy replacement by unemployed or marginal workers).
However, this categorisation of strategies ignores potential

56

changes in the labour process which may effect some homogenisation of labour market conditions; and transformations
within the capitalist economy which ensure that some of
these strategies of segmentation are spatially differentiated.

An example which illustrates the latter point has been
the growth of productive decentralisation, especially in
Italy, but also in the UK (Cooke 1983). This involves the
sub-contracting of parts of the production process to small,
often family firms located in predominantly rural areas
away from centres of strong labour market demarcation and
solidarism. This decentralisation both cheapens the labour
input through the employment of the substantial pools of
the 'green' rural labour reserve, and it disciplines the
remaining labour force through a policy of spatial diversification. Cooke argues that developments in Central Wales
are a more general form of this: he describes the area as
"a state-managed branch plant outpost of the midlands and
south of England" (1981, p. 25). Manufacturing industry,
especially of relatively small, sub-assembly and components
industries, has greatly expanded in recent years because of
the existence of agricultural labour reserves, low levels
of unionisation (in comparison with demarcation and exclusion strategies elsewhere), and the encouragement of the
Development Board for Rural Wales. This has produced a
locally-distinct social structure based on an externally-
located capitalist class, a substantial petty bourgeoisie
and a growing rural proletariat. In other words, the rapid
expansion of manufacturing industry in central Wales
(313 per cent increase, 1959-75) represents a partial
decomposition of the industrial classes of modern capitalism.
Capital has been simultaneously centralised and inter-
nationalized as it rotates in and out of different areas,
while labour's patterns of demarcation and exclusion have
been partly decomposed as it has become at least partly
ruralized, both in the Third World and Britain.

This analysis is suggestive of other points about rural
social relations. First, the consequences of such developments may well produce social polarisation within 'rural'
areas. As Davies says:

> the introduction of industry allows
> particular sections of the local
> society to jump on the bandwagon
> represented by the industry, notably,
> sectors of the working class such
> as skilled workers, especially those
> near the area, and those sectors of
> the petty bourgeoisie whose capital
> is invested in retail consumption.
> A large percentage of the population
> will not participate, will have their
> relative life chances reduced . . . as
> they move further down the queue for
> the scarce social infrastructure which
> exists in the area. (1978, p. 96)

Second, when it is asserted that there are substantial increases in the number of managerial and professional workers in rural areas, it should be specified what kinds of labour market qualifications they possess - whether these are 'inclusive' or 'general'. It should also be determined how these workers are related to the functional division of labour characterising the sectors involved, identifying in particular, whether they function as a 'service class' for capital, or as part of the state, or as relatively 'deskilled white-collar workers' (see Abercrombie and Urry 1983 on these distinctions); and how they are related to the pertinent *spatial* divisions of labour, and attendant forms of labour market segmentation, especially whether they function within central, semi-peripheral or peripheral plants.

Third, in order to unravel the socio-political consequences of these processes it is important to have some understanding of political struggles prior to the recent periods of restructuring and economic decline. Broadly speaking, the most significant form of oppositional struggle within the industrial period of British capitalism was economic militancy, combined with support for separate political struggle within the Labour Party. This pattern was found in the major urban-based industries - coal, steel, docks, railways, engineering, automobiles etc. In each case there were a number of distinctive features: large numbers at each workplace, a high proportion of male workers, some development of an occupational community, and the centrality of that industry to the national economy. Yet, at the same time, many areas both urban and rural were not economically militant. It is now necessary to consider not only what are the forms of politics typical of an economy experiencing profound restructuring and decline but also what effects this pattern will have on *existing* forms of political organisation.

Massey (1983) contrasts the position in south Wales with that in Cornwall. In the former it is not difficult to see how a number of conditions ensured the pattern of economic militancy: a relatively undifferentiated working class; a single union; pride in the masculine character of mining and steel work; the relative lack of a new middle class or of small entrepreneurial capital; and the lack of alternative forms of labour. However, the decline in the mining industry and the arrival of multinational plants and new forms of service employment threaten the power and dominance especially of the mineworkers. There are the following effects: an increase in the size of the 'new middle class' and thus a blurring of lines of conflict between labour and capital; greater 'external control' of the region, increasing the difference between south Wales and more 'central' regions and the demand for more top management and professional jobs *within* the region (rather than the regaining of the functions of conception and control); a decrease in the average skill level and an increase in semi and unskilled labour especially for women, thereby undermining the homogeneity, uniqueness, maleness, income

and status of the previous dominant forms of labour; and
the introduction of capitalist or state wage-relations to
a relatively inexperienced and unorganised labour force.
In contrast, in Cornwall the impact of similar industrial
and occupational changes has had the following, opposite
effects: an increase in waged labour, not a depression of
wage levels; a more homogeneous working class and other
waged-sectors; and a threat of competition to traditional
capital in both the labour and commodity markets, thus
focussing previously highly blurred lines of conflict
between capital and labour. Hence, the impact of a roughly
similar pattern of accumulation and state activity seems to
produce different effects because it is articulated with a
quite different pre-existing structure. Massey convincingly
shows that in order to assess the significance of certain
economic trends, we cannot characterise areas simply in
terms of present industrial, occupational or class changes.

CONCLUSION

In this chapter, I have attempted to demonstrate that
various critical notions - of different, overlapping spatial
divisions of labour, of all localities as sites for the
reproduction of labour-power, of variations in local social
structures, etc. - render problematic the notion that there
are distinct 'rural' localities. The 'new international
division of labour' involves not just the export of industrial employment to rural localities in the Third World -
but also, to some extent, to such locations in the first
world. This is highly variable, though, and will not ensure
that there is a distinct rural social structure. However,
although the effects may even be heightened class relations
within certain rural areas, the overall consequence must
be to undermine important social bases for class actions.

    The effects of class recomposition can be further
seen by stating fairly formally the conditions under which
the actions of individuals are more likely to take a class
character (cf. Elster 1978). They are more likely to do so
the more:

1. The spatially separated experiences of groups of people
can be interpreted as the experiences of a whole class.
This depends upon particular local civil societies being
both class-divided and perceived as similarly structured by
class rather than by other significant social entities.
The recent developments in local social structures and the
variable role of agrarian class relations makes this an
increasingly difficult condition to meet.

2. There is a high rate of participation and concerted
action within a range of spatially specific but overlapping
collective organisations. However, the recomposition of
classes has undermined the labour market strategies of
'demarcation' and 'solidarism' which sustained overlapping
collective organisations.

3. Other collectivities within local civil societies are organised in forms consistent with that of class rather than being in conflict with it. Organisations which seek to re-capitalise particular localities will have difficulty in not reinforcing conflicts of locality versus locality, especially with 'exposure' as a significant form of labour market strategy. There is a heightened probability of a 'local corporatism' in which diverse classes come together to heighten the distinctiveness of each locality as a pool of wage-labour (Hernes and Selvik 1981).

4. Other kinds of gains and benefits which could be attained through non-class actions (such as higher incomes, lower prices, better conditions of work and leisure, etc.) are perceived to be, and are unavailable. This will be more likely where social inequalities are believed to be produced by antagonistically structured *national* or at least *regional* classes. Again the growth in locally based systems of social stratification make this condition diff-icult to sustain at the same time that there are a wide variety of alternative social groupings which can obtain certain gains and benefits.

5. Large numbers of individuals in different spatial locations conclude that class actions can be successful. However, the spatial recomposition of classes, in part in green-field sites, together with high rates of geographical mobility to escape from class-divided industrial urban localities, also makes this condition difficult to fulfil.

Changes in the structuring of certain contemporary capitalist societies are such that new concepts are necessary in order to make sense of what I term 'former industrial countries', of which the UK is the leading example. Undoubtedly the changing relations between formerly urban-industrial and rural-agricultural areas are one aspect of this restructuring and the recomposition of the industrial classes of such countries. In this context, it is not only rural sociology which is losing its distinctive focus but also the categories and concepts applicable to theorising the 'urban' and the 'industrial'.

NOTES

[1] See my elaboration of the 'economy/civil society/state' formulation in 1981a and 1984.

REFERENCES

Abercrombie, N. and Urry, J., 1983. *Capital, Labour and the Middle Classes,* (London: Allen and Unwin).

Aglietta, M., 1979. *A Theory of Capitalist Regulation,* (London: New Left Books).

Banaji, J., 1976. Summary of selected parts of Kautsky's *The Agrarian Question, Economy and Society,* 5, 2-49.

Bhaskar, R., 1979. *The Possibility of Naturalism,* (Brighton: Harvester).

Broadbent, T.A., 1977. *Planning and Profit in the Urban Economy,* (London: Methuen).

Carney, J., Hudson R., Lewis J. (eds.), 1980. *Regions in Crisis,* (London: Croom Helm).

Cooke, P., 1981. *Local Class Structure in Wales,* (Cardiff: Dept. of Town Planning, UWIST, Papers in Planning Research no. 31).

Cooke, P., 1983. *Theories of Planning and Spatial Development,* (London: Methuen).

Davies, T.M. 1978. Capital, state and sparse populations: the context of further research, in: H. Newby (ed.) *International Perspectives in Rural Sociology,* (Chichester: John Wiley), 87-104.

Donnison, D. and Soto, P., 1980. *The Good City,* (London: Heinemann).

Dunford, M., Geddes, M. and Perrons, D., 1980. *Regional Policy and the Crisis in the UK: A Long-Run Perspective,* (London: City of London Polytechnic, Department of Economics and Banking, Working Paper 3).

Dunleavy, P., 1980. *Urban Political Analysis,* (London: Macmillan).

Elster, J., 1978. *Logic and Society,* (Chichester: Wiley).

Fothergill, S. and Gudgin, G., 1979. Regional employment change: a sub-regional explanation, *Progress in Planning,* 12, 155-219.

Fothergill, S. and Gudgin, G., 1982. *Unequal Growth,* (London: Heinemann).

Gasson, R., 1966. Part-time farmers in south-east England, *Farm Environment,* 11, 135-139.

Gasson, R., 1980. Roles of farm women in England, *Sociologia Ruralis,* 20, 165-180.

Gershuny, J., 1978. *After Industrial Society? The Emerging Self-Service Economy,* (London: Macmillan).

Grigg, D., 1969. Regions, models, and classes, in: R.J. Chorley and P. Haggett (eds.) *Integrated Models in Geography,* (London: Methuen), 461-509.

Harris, R., 1983. Space and class: A critique of Urry, *International Journal of Urban and Regional Research,* 6, 115-121.

Hernes, G. and Selvik, A., 1981. Local corporatism, in: S. Berger (ed.) *Organizing Interests in Western Europe: Pluralism, Corporatism, and the Transformation of Politics,* (Cambridge: Cambridge University Press), 103-119.

Humphries, J., 1977. Class struggle and the persistence of the working class family, *Cambridge Journal of Economics,* 1, 241-258.

Josling, T.E., 1974. Agricultural policies in developed countries: a review, *Journal of Agricultural Economics,* 25, 229-263.

Keeble, D., 1976. *Industrial Location and Planning in the UK,* (London: Methuen).

Keat, R. and Urry, J., 1982. *Social Theory as Science,* 2nd edition. (London: Routledge & Kegan Paul).

Kennett, S., 1982. Migration between British local labour markets and some speculation on policy options for influencing population distributions. *British Society for Populational Studies Occasional Paper 28.* (Conference on Population Change and Regional Labour Markets, OPCS), 35-54.

Kreckel, R., 1980. Unequal opportunity structures and labour market segmentation, *Sociology,* 14, 525-550.

Lebowitz, M., 1980. *Capital as finite,* (paper given at the Marx Conference, University of Victoria, Canada).

Lipietz, A., 1980. The structuration of space, the problem of land and spatial policy, in: J. Carney *et al.* (eds.) *Regions in Crisis,* (London: Croom Helm), 60-92.

Massey, D., 1978. Regionalism: some current issues, *Capital and Class,* 6, 106-125.

Massey, D., 1981. The UK electrical engineering and electronics industries: the implication of the crisis for the restructuring of capital and locational change, in: M. Dear and A.J. Scott (eds.) *Urbanization and Urban Planning in Capitalist Society,* (London: Methuen), 199-230.

Massey, D., 1983. Industrial restructuring as class restructuring, *Regional Studies,* 17, 73-89.

Massey, D. and Meegan, R., 1982. *The Anatomy of Job Loss,* (London: Methuen).

Murgatroyd, L. and Urry, J., 1983. The restructuring of a local economy: the case of Lancaster, in: J. Anderson *et al.* (eds.) *Redundant Spaces,* (London: Academic Press).

Newby, H., 1977. *The Deferential Worker,* (London: Allen Lane).

Newby, H. , 1978. The rural sociology of advanced capitalist societies, in: H. Newby (ed.) *International Perspectives in Rural Sociology,* (New York: Wiley), 3-30.

Newby, H. , 1979. *Green and Pleasant Land?,* (London: Hutchinson).

Newby, H. , 1980a. Rural sociology, *Current Sociology,* 28, 1-141.

Newby, H. , 1980b. Urbanization and the rural class structure: Reflections on a case study, in: F. Buttel and H. Newby (eds.) *The Rural Sociology of the Advanced Societies: Critical Perspectives,* (London: Croom Helm,), 255-279.

Newby, H. , 1982. Rural sociology and its relevance to the agricultural economist, *Journal of Agricultural Economics,* 33, 125-165.

Newby, H. *et al,* 1978. *Property, Paternalism and Power,* (London: Hutchinson).

Pahl, R. , 1965 *Urbs in Rure,* (London: London School of Economics, Geographical Papers, no. 2).

Pahl, R. , 1968. The rural-urban continuum, in: R. Pahl (ed.) *Readings in Urban Sociology,* (Oxford: Pergamon).

Pahl, R. , 1980. Employment, work and the domestic division of labour, *International Journal of Urban and Regional Research,* 4, 1-20.

Saunders, P. , 1980. *Urban Politics,* (Harmondsworth: Penguin).

Saunders, P. , 1982. *Social Theory and the Urban Question,* (London: Hutchinson).

Saville, J. , 1957. *Rural Depopulation in the United Kingdom,* (London: Routledge & Kegan Paul).

Sayer, A. , 1982. Explanation in economic geography: abstraction versus generalisation, *Human Geography,* 6, 68-88.

Stinchcombe, A.L. , 1961. Agricultural enterprise and rural class relations, *American Journal of Sociology,* 67, 169-176.

Storper, M. , 1981. Towards a structural theory of industrial location, in: J. Rees *et al* (eds.) *Industrial Location and Regional Systems,* (London: Croom Helm).

Urry, J. , 1981. *The Anatomy of Capitalist Societies,* (London: Macmillan).

Urry, J. , 1981b. Localities, regions and social class, *International Journal of Urban and Regional Research,* 5, 455-474.

Urry, J. , 1983a. De-industrialisation, classes and politics, in: R. King (ed.) *Capital and Politics* (London: Routledge & Kegan Paul).

Urry, J., 1983b. Some notes on realism and the analysis of space, *International Journal of Urban and Regional Research,* 5, 122-127.

Urry, J., 1984. Space, time and social relations, in: D. Gregory and J. Urry (eds.) *Social Relations and Spatial Structures,* (London: Macmillan).

Walker, R., 1978. Two sources of uneven development under advanced capitalism: spatial differentation and capital mobility, *Review of Radical Political Economics,* 10, 28-37.

Walker, R. and Storper, M., 1981. Capital and industrial location, *Progress in Human Geography,* 5, 473-508.

Westaway, J., 1974. The spatial hierarchy of business organisations and its implication for the British urban system, *Regional Studies,* 8, 145-155.

4. Segmentation in local labour markets

TONY BRADLEY

Britain is a remarkably uneven and unequal society.
The social divisions between rich and poor, lofty and lowly,
powerful and powerless are, if anything, widening during the
current era of monetarist political economy and mass
unemployment. For all but a tiny minority of capital-
owners, the dominant reward structure continues to be tied
to the organisation and allocation of work. Some groups of
workers, however, have notably greater job security than
others, are more affluent, enjoy better working conditions,
and are more effectively organised to protect their interests.
This chapter examines such divisions in work opportunities
and rewards arising from the differentiated structure of
rural labour markets. As the evidence presented demonstrates,
many rural workers are low paid, suffer poor work conditions,
and are submissive in the face of their employers' power.
This raises the question: why has the *rural* working class
remained quiescent in contrast to the general economic
militancy at work, and the political mobilisation of other
marginal groups, in cities throughout Britain? My intention
is to provide some answers to this question, through an
examination of local labour markets in rural England.

MODELS OF LABOUR MARKET SEGMENTATION

Various theories have emerged which challenge the classical
premise that wage rates and markets for labour conform to
equilibrating tendencies, arising from the interaction of
forces of supply and demand within a single, universal,
national labour market. Since job loss, economic decline
and de-industrialisation have not been uniform within the
British labour market, then, it seems logical to look for
other influences, in addition to the simple interplay of
market forces, in the restructuring of labour power into
its constituent labour markets. Analysis is required to
explain why certain industries, localities, and occupational
and social groups have fared better than others. The
crucial question is to what extent and how are labour markets
segregated from each other.

Explanations for the uneven distribution of employment generally follow either 'dual' or 'segmented labour market' hypotheses. Simply stated, the 'dualists' (Bluestone 1970; Doeringer and Piore 1971) assert that labour markets in advanced industrial societies have developed a dichotomous structure. On the one hand, there is a primary sector characterised by high wages and fringe benefits, job security, opportunities for career mobility, high status, work variability and satisfaction, self-pacing, on-the-job training, and high levels of unionisation. On the other hand, there is a secondary sector characterised by low pay, instability, lack of job prospects, low occupational status, routine, low levels of job satisfaction, lack of training, and non-unionisation. The basic conclusion drawn from various American studies is that certain groups of workers - chiefly women and ethnic minorities - even when able to enter the labour market, are confined to low-paid, insecure and dirty ('shit-work') jobs (see Bosanquet and Doeringer 1973 for parallel British findings).

More sophisticated models of labour market structure emphasise the differential access of workers to *segmented* labout markets (SLMs), according to various criteria: including the power of corporations to control 'internal' labour markets (Doeringer and Piore 1971); the degree to which particular work is routinised and disciplined (Edwards *et al*. 1975); occupation; age and sex. A number of criticisms of both dual and segmented labour market hypotheses have been raised. The most general question their reliance on economic and industrial structure, maintaining that historical accounts of segmentation are overly condensed (Kreckel 1980); the data used is circumstantial (Blackburn and Mann 1979); the models developed are descriptive and not explanatory (Townsend 1979, Levitan *et al*. 1976); and the accounts given rely on functionalist assumptions about the ability of capitalists to coercively control the labour force (Abercrombie and Urry 1983).

It is in this context that a number of sociological treatments have attempted to inject an action frame of reference into labour market debates, by connecting the idea of SLMs to theories of social stratification (Offe and Hinrichs 1977; Blackburn and Mann 1979; Kreckel 1980; Abercrombie and Urry 1983). Each adopts a broadly similar standpoint: that the access of employees to particular occupations in specific labour markets is primarily conditioned by social relations of power, established within the work situation, as negotiated through the labour process. As such, segmentation results from the interactional opposition of class powers, rather than from the structural conditions of differentiated markets for labour. Thus Blackburn and Mann identify the principal obstacle to the development of a united working class as "the development of the forces of production themselves":

> As capitalism has expanded throughout
> the world, new groups have been brought

into the labour market, possessing
differing values and skills from those
already there. The assimilation of
rural immigrants, women, slaves and
international migrants into the existing working class has been slow. Each
group has at first constituted a kind
of *lumpenproletariat,* or secondary
labour force. It has been in the
immediate economic interests of the
primary work force to keep them in that
position.
(Blackburn and Mann 1979, p. 33)

In other words, action to change their position within
the labour market is partially in the hands of the workers
themselves, but will be constrained, not only by the actions
of employers, but also by the relative power of other
working groups. Thus, for example, processes of profess-
ionalisation and educational inflation condition changes
in long-term patterns of social mobility and stratification
(Johnson 1977; Larson 1977; Halsey *et al* 1980) which benefit
one group of workers - the service class - to the detriment
of others. The *power* (market capacity variable is, therefore,
of crucial importance in explaining the development of SLMs.
Both dual and segmental descriptions of labour markets
neglect the origins of social stratification *within* the
working class. The labour process theories, on the other
hand, give a reason for the emergence of barriers between
segregated labour markets, namely, the asymmetrical relations
of power between workers engaged in differentiated labour
processes.

If the question of *social* segregation within labour
markets is theoretically confused, the divisions between
labour markets on the basis of *geography* is equally contested,
on methodological grounds (Robinson 1967). Nevertheless,
all *local* labour market (LLM) definitions share one common
feature, which is the attempt to specify boundaries between
'functional areas', according to economic activity. As
Hunter and Reid (1968, p. 42) put it; "The essential points
about a local labour market are that the bulk of the area's
population habitually seeks employment there and that local
employers recruit most of their labour from that area."
In practice, empirical investigations have tended to
concentrate on the labour *supply* side, although economists'
definitions of LLM stress the monopsonistic powers of
individual firms (see Metcalf and Richardson 1972) in
recruiting employees from a clearly defined locality.

Certain difficulties surround the LLM concept when
applied to rural areas. The most obvious of these is the
sheer spatial dispersal of the working population, and
diversity of firms, across any particular territory. Any
area selected, on the basis of occupational category,
employment type and industrial group, as a 'self-contained'
LLM will include a wide variety of individual sub-markets.

Rural LLMs are, in Clarke Kerr's term (1954), balkanised. In consequence, policy-makers have usually avoided the difficulty of specifying rural LLMs by arbitrarily including these sub-regions within spatially-large, urban labour markets, on the assumption that the more peripheral localities supply 'commuters' into employment centres. This appears to be the reasoning behind the Department of Employment's 'travel-to-work' areas (see Smart 1974). Even so there are sound reasons for not dispensing with the notion of rural LLMs.

First, certain industrial sectors (agriculture, fisheries, forestry, tourism) and occupational groups (farmer, farmworker, share-fisherman) are under-represented in non-rural areas. Even so, rural LLMs are not exclusively defined by these industrial and occupational categories. Within the sociological and labour history literatures, however, the 'industrial' workforce of the rural areas is treated as almost entirely composed of agricultural workers (Newby 1977; 1980). Rural poverty and deprivation is, thus, elided with agricultural work conditions (Giles and Cowie 1957; Minchinton 1968; Newby 1977; Brown and Winyard 1979; Winyard 1982), even though low pay on the farm is neither the only nor indeed the primary outcome of the unequal structure of incomes amongst the rural population.

Second, the geographical size of travel-to-work areas means that they subsume and conceal a rich variety of local conditions. Their basic weakness is the statistical assumption that the modal range of commuting distances falls between the median and the maximum values, when one or several LLMs may be embedded within the functional area.

Third, there is a social bias in the specification of LLMs, according to travel-to-work areas, resulting from the emphasis placed on patterns of commuting. Such a ready acceptance of the separation of work and home localities inevitably leads to the formulation of policies according to the everyday experiences of socially and geographically mobile strata within the local (residential) population.

The final reason for distinguishing rural labour markets is that they seem to have undergone considerable changes in recent years. As Fothergill and Gudgin (1982) have shown, the small market towns in England grew more rapidly than any other settlement group during the 1970s, in terms of new manufacturing employment. Furthermore, there is circumstantial evidence to suggest that the rural employment centres fared relatively well in recording proportionately lower levels of job loss than larger centres over the same period (Townsend 1981).

Surprisingly little research has been conducted into the structure of rural LLMs. Apart from the tangential coverage of various community studies (e.g. Nalson 1968) three investigations have specifically addressed the question (Hale 1971; Wenger 1980; McDermott and Dench 1983). In each of these surveys - all conducted in Wales and the west of England - the issue at stake has been local employment opportunities for rural youth. The conclusion, common to each study, is that the immediate prospects for young people are bleak unless they move away. The narrow focus of these studies, however, precludes any general conclusions about the differentiation of labour markets.

The data reported here is taken from a study of deprivation in rural areas funded by the Department of Environment. This study aimed to provide evidence on the styles of living and conditions of existence of the pop- ulation *resident* in rural England. Largely quantitative information was collected during 1981 from 870 households living in five English localities, relevent to the issue of *social* rather than 'accessibility' or 'opportunity' deprivation (Shaw 1979; Moseley 1980), and included data, from all adult household members, on labour market positions.

Random sample surveys were conducted in the five areas, which were chosen as representative of various 'ideal-type' rural labour markets. These were:

1. A metropolitan labour market in north-west Essex.

2. An area of upland family farming in the Yorkshire Dales National Park.

3. An area of capital-intensive agriculture in mid-Suffolk.

4. A 'transitional' area of family farming and commuting (to the West Midlands) in south-west Shropshire.

5. A labour market divided between tourism and share- fishing on the Northumberland coast.

Each of the study areas lies within one of the travel- to-work areas defined by the Department of Employment (Smart 1974; Gillespie 1977). In a strict sense they are not LLMs but labour residence areas. The most localised labour markets are embedded within them, although commuting labour markets extend beyond the boundaries of the study areas. They are, therefore, 'functional' for those workers who either choose, or are constrained, to work in the immediate locality of their residence.

To make maximum use of the employment information from our household samples the data have been divided into four groups of variables. These component groups are:

1. Structural: employment sector, socio-economic group, occupation grade, size-of workforce, earnings and wage rates.

2. Social Group: socio-economic group (economic class), locals/non-locals, sex.

3. Market Capacity (Power): education, qualifications, labour market 'mode of entry', unionisation.

4. Deprivation: low pay, scale of 'deprivation at work'.

The structure of the LLMs is given by simple frequency distributions of the basic structural variables. These variables are, therefore, to be regarded as independent, with the exception of socio-economic group (S-EG) which, for our purposes, is used as a measure of economic class. The population has been split into sections along the axes of class, sex and 'ethnicity'.

It is widely accepted that sex and occupation fundamentally divide collective labour power into segregated working groups. Similarly, ethnicity is recognised as a fulcrum around which labour market segmentation revolves, particularly between blacks and whites in urban labour markets. Here, ethnicity is defined as 'localness'. Geographical mobility has been shown to be a major source of segregation in access to domestic property in rural areas (Shucksmith 1981; Winter 1980). Consequently, I have attempted to examine the evidence for labour market segregation between locals and non-locals in our five rural areas.[1]

Having located various occupational classes, men and women, locals and non-locals in the structure of rural LLMs I turn to the 'causal power' (Abercrombie and Urry 1983) which each social group has at its disposal, and the degree of deprivation at work which it faces.

STRUCTURE OF THE LOCAL LABOUR MARKETS

The structure of LLMs may be specified in various ways. Whereas agriculture is the dominant industrial *sector* in three of our areas - Yorkshire Dales, Shropshire and Suffolk - only in the last is the social division of labour market-based in capitalist production. Even here social relations in the work-place are much modified from those which pertain in other 'industrial' work situations (Newby 1977). In the Dales and Shropshire, the agrarian social structure is dependent upon the use of family labour, under conditions of 'simple commodity production'. Although paid similar wages to non-family workers, in the short term, the prospect of future ownership, or at least occupation of the holding, crucially improves the life chances of these family wage labourers. Nor are these familial relations of production confined to agriculture. Share-fishing, which occupies a substantial proportion (15 per cent) of the Northumberland coast labour force and is the single most important sector, numerically, is also structured around the employment of 'propertied labour' (Davis 1980).

Other sectors are less frequently represented in terms of economic activity but contribute more labour *market*

70

positions. The metropolitan labour market is characterised by white-collar work. One-quarter of all economically active individuals work in general management and the professions, with an additional 14 per cent 'supporting' these grades in secretarial and clerical work. Significantly, the proportion of the employed population engaged in *public* sector professions - doctors, school-teachers, nurses - was constant for all areas (8 per cent), reflecting the state's concern for territorial justice. In the two areas most dependend upon tourism - the Dales and Northumberland coast - 30 per cent of employees were involved in shop work (retailing) and personal services (including catering, cleaning and hairdressing), compared with 20 per cent in the other three areas. Taken together, small building firms (construction) and haulage contractors (transport) occupy a constant 10 per cent of the labour force.

Sectoral statistics confuse the effects of social and technical divisions of labour. Social differentiation is more clearly revealed, therefore, by reconstructing the occupational class structure according to socio-economic group (Table 4.1). The contrast is most striking between the Essex and Northumberland coast LLMs. Whereas 66 per cent of the Essex sample are occupied within the upper three S-EGs; some 64 per cent of the Northumberland coast sample are within the lower three S-EGs, indicating wide divergences in local class structures. The balkanisation of rural LLMs is underlined by the fact that only three specific occupational grades - farmers, agricultural workers and non-manual workers - included more than 5 per cent of the workforce in each LLM. Furthermore, each of these grades represents a wide variety of specific occupational positions.

Size of workforce and place of work are important indicators of LLM structure. Both these variables reinforce the picture of a small-scale, fragmented labour force. In dramatic contrast to conventional images of labour markets in advanced industrial societies, over one-quarter (26 per cent) of the economically active in the five LLMs worked at home. Naturally, the proportion was greatest for employers and managers (60 per cent of S-EG2) - including the various types of farmer. More importantly, perhaps, only 70 per cent of persons occupied in skilled manual grades worked away from home. Additionally, two-thirds (65 per cent) of all the economically active were working in labour forces of less than ten workers. This ranged from as few as 42 per cent in the metropolitan labour force (reflecting the presence of considerable numbers of London commuters) to as many as 79 per cent of the Yorkshire Dales sample. Clearly, the *vast* majority of the rural labour force, within and outside agriculture, work in *very* small units. It is likely that many of those working outside of agriculture experience similar paternalistic modes of control as those that Newby (1977) describes confronting the agricultural workforce.

Table 4.1   Proportions of all economically active
             individuals in various socio-economic groups,
             by area

General House-hold Survey socio-economic groupings	Essex	Yorkshire	Suffolk	Shropshire	Northum-berland
1. Professional	19.8	11.8	14.2	16.7	9.3
2. Employers and managers	22.4	25.0	19.8	26.5	9.8
3. Intermediate and junior non-manual	23.7	14.1	14.2	14.5	17.1
4. Skilled manual	15.5	26.9	18.3	19.2	33.3
5. Semi-skilled manual and personal service	13.8	17.9	27.7	20.9	18.1
6. Unskilled manual	4.7	4.2	5.6	2.1	12.2
	N = 232	N = 212	N = 267	N = 234	N = 204

Source: Author's survey, 1981

The final structural indicator I wish to use is the
distribution of *average gross weekly earnings* (Table 4.2).
The ranking of earnings for the rural employment groups is
the same as for the country as a whole: male, full-time
non-manual workers receive highest earnings; and female,
part-time manual workers the least. Relative inequality
is, however, greater in the rural areas. In absolute
terms, the people at the top are even better off than in
the rest of the country, whereas the plight of those at
the bottom is more desperate. The explanation for the
affluence of male non-manual groups is that they do not
work in the most localised labour markets but are commuters
to large employment centres. Herein we can see the articulation of powers between LLMs and markets in domestic
property. The relatively advantaged position that these
workers face in the labour market, usually sustained by the
use of qualification-specific exclusion practices, gives
them (and their families) greater market capacity in the
county 'house-and-garden' stakes. Significant local
differences also emerge. Average male manual wage-rates
are 20 per cent lower in the Yorkshire Dales and Northumberland coast localities than the other rural areas. In
contrast, part-time women's manual labour receives highest
rewards in Northumberland, because of a highly competitive
tourist industry.

The labour market structure of rural England is, thus,
characterised by large numbers of firms employing small
labour forces. Capital ownership is predominantly controlled
by a rural petit-borgeoisie, able and willing to pay very
low wages.

CLASS, SEX AND ETHNIC ASYMMETRIES OF POWER

Class, sex and ethnic groupings were compared according to
their power to exploit labour market opportunities and,
hence, to influence patterns of social differentiation.
Two pairs of variables were used as indicators of market
capacity. The first pair - years of full-time education
and final qualifications - represent the acquisition and
possession of achievement-related entry requirements.· The
second pair - sources of employment advice and unionisation -
reflect the individual's capacity to use other institutional
resources for labour market entry.

Certain general variations in levels of skill were
discernible between the five local populations. The Essex
(metropolitan) and Shropshire (transitional) areas had
relatively highly educated labour forces, with over one-
quarter having 14 or more years education and far fewer
unqualified workers (35 per cent). In the other three
areas 70 per cent of the adult population had left school
at the minimum age, with 50 per cent of the economically
active having no qualifications. Such differences in LLM
skilling reflect the presence or absence of a significant
professional service class within the local social structure
(Table 4.1).

Table 4.2   Average gross weekly earnings (£) for various employment groups by area

	Essex	Yorkshire	Suffolk	Shropshire	Northumberland	All areas	GB
Full-time male non-manual	205.6	206.3	201.4	291.5	159.3	215.2	163.1
Full-time male manual	121.6	103.3	125.8	131.3	102.8	117.2	121.9
Full-time female non-manual	108.3	113.1	119.2	96.9	83.1	104.1	96.7
Full-time female manual	48.2	69.7	78.3	71.4	56.8	66.3	74.5
Part-time female non-manual	42.7	52.8	48.7	57.1	45.2	48.74	43.5
Part-time female manual	27.2	20.9	22.5	15.7	27.5	21.7	33.0

Source: Author's survey and New Earnings Survey, 1981

Staying on at school is positively correlated with occupational class position (S-EG). Whereas 55 per cent of professionals stayed on after 16, only 22 per cent of skilled manual and 3 per cent of unskilled manual workers did the same. There is an evident divergence from this neat linear progression, however, in respect of employers and managers (including farmers) and semi-skilled manual grades (including agricultural workers). Identical proportions (55 per cent) of these antagonistic class groupings finished their schooling before the age of 16. One explanation is that work orientations in agriculture are mediated by an 'ideology of experience' wherein qualifications are shunned as a key to entry and progression through the local labour market (Bradley 1979). Farming and other local work is strongly associated with practical skills which can, it is argued, only be gained experientially. Of course, the acceptance by rural workers of such an ideology severely restricts their movement into other, non-local labour markets and increases their dependence upon immediate employers. These self-same members of the petit-bourgeoisie need not concern themselves with occupational mobility, except in those cases where there is a barrier to inter-generational acquisition of family property.

This 'ideology of experience' is even more evident when locals and non-locals, and men and women are compared over their years of schooling. Twice as many 'newcomers' as 'natives' stayed on after 17. Many of the former will have undergone a process of educational socialisation which emphasised achievement and the competitive acquisition of formal qualifications. The women in the rural LLMs were also more highly educated than the men, with double the proportion of women as men (13.5 compared with 6.4 per cent) having attained 'O' level standard. Arguably, though, women are 'educated out' because they have fewer opportunities to take hold of family property rights than men; or because they are largely confined to lower status occupations in which experience counts much more than formal qualifications.

A large proportion of workers in each of the study areas used particularistic sources of advice - friends, relatives, personal contacts - as a means to a job. In Suffolk, Shropshire and Northumberland only one-fifth of workers relied on more formal 'advice' - from school, job centre, national and local media - in getting their current work. Even fewer used these channels in the Yorkshire Dales, whereas in Essex, as would be expected from a more 'skilled' workforce, one-third of the sample consulted formal services, particularly the job centre and national media. The appropriation of these specific institutional resources does not, however, appear to be class-dependent. Although larger proportions of the lower S-EGs cited the use of informal advice over half of the 'top people' also relied on particularistic contacts. This is not as surprising as it may seem. It is quite likely that the use of family contacts and friendship links by professionals represents the operation of inter-generational 'cycles of advantage' (Halsey *et al*. 1980). We may all derive a benefit from family and friends but, of course, the more influential *they* are the 'better' *we* can do.

The family background of workers - whether local or non-local - was strongly correlated with the use of different sources of advice. Twice as many of the non-locals used formal channels, whereas local people relied far more heavily on family, friends and kinship contacts for securing a job. Even so, a substantial minority of locals (more than one-quarter) were reliant on the local press and formal, institutional contacts for their jobs. Two-thirds of this group were located in the lower three S-EGs, indicating that segregation exists in these LLMs *between* locals, on the basis of their differential power to exploit localistic ties. Arguably, these people are the least secure fraction of the rural working class.

Unionisation and affiliation to other associations of social closure constituted the fourth power resource available to rural agents. None of the five LLMs were more than one-third unionised. In general it may be said that the rural workforce is *extremely* thinly unionised. Perhaps unionisation is dependent upon occupational class position and local ethnicity in rural labour markets? The evidence presented in Table 4.3 demonstrates a very definite connection between unionisation and class. But, the picture that emerges is startlingly counter-intuitive. Levels of unionisation are directly related to and, quite possibly, conditioned by how close an individual is to *the top* of the class structure. The top two groups each have levels of affiliation (unions and associations taken together) greater than 50 per cent. The particularly high figures for S-EG2 reflect the strength of the National Farmers' Union in recruiting all types of farmers. The NFU would not, however, be expected to engage in radical action in support of the rural working class.

In writing on the National Union of Agricultural and Allied Workers, Newby has commented:

> Compared with many of its urban industrial counterparts the NUAAW therefore remains a weak union - and a poor one. It has neither the power nor the resources to counter the pressure of the exceedingly well-organised farmers' lobby, nor is the level of unionisation in agriculture sufficiently high - it is no more than 40 per cent - to permit the contemplation of widespread militant activity (Newby 1980, p. 143).

And elsewhere:

Table 4.3 Proportions of individuals in various socio-
economic groups who are or are not affiliated
to a union or other professional association

General House-hold Survey socio-economic groupings	Not affiliated	Member of union	Member of association	Missing data
1. Professional	44.9	32.3	21.5	1.3
2. Employers and managers	42.1	41.2	10.4	6.3
3. Intermed- iate and junior non-manual	69.8	22.9	2.6	6.4
4. Skilled manual	63.1	22.0	6.7	7.5
5. Semi- skilled manual and personal service	68.7	24.8	0.7	5.8
6. Unskilled manual	81.5	18.5	–	1.0

Source: Author's survey, 1981

> Hence the NUAAW has become trapped
> within a vicious circle - its lack
> of power is based ultimately upon a
> low level of organisation which in
> turn stems partly from its weakness
> in wage bargaining
> (Newby 1977, p. 256).

Nevertheless, 40 per cent unionisation amongst agricultural workers sets them apart as well represented compared to the mass of rural workers.

In the case of unskilled manual workers in these peripheral areas, fewer than one-fifth are affiliated to any organisation which is willing (whether effective or not) to represent their interests and needs within the sphere of economic bargaining over pay and work conditions. Those who are, arguably, most in need receive the least benefit from the traditional organs of working class militancy. Non-affiliation is partly conditioned by the petit-bourgeois structure of the LLMs and the extreme resistance of rural employers to the intervention of unions. Also linked, however, is the ambivalence, apathy and recalcitrance of the unions, particularly the larger ones, towards the rural worker whom they see as remote, isolated, antagonistic towards 'the idea of unions', and lacking in industrial muscle.

These extremely low levels of unionisation pose a very specific problem for the rural working class. The inability to call upon wider sections of the labour movement in order to protect wage rates, mobilise for adequate working conditions, or enforce calls for industrial action, implies a degree of marginalisation felt by few other groups of British workers, with the exception of women and ethnic minorities. In general, the rural working class is relatively low-skilled, dependent on personal and particularistic ties to petit-bourgeois and family employers and thinly unionised.

The structural importance of the farming sector should not be undervalued as a factor conditioning the overall shape of the rural labour process. Agriculture, and particularly capitalist agriculture, exerts a powerful influence, as other chapters demonstrate, over the politics of the rural areas and definitions of what constitutes legitimate action. Ironically, the capitalist farmer's hegemony derives from models of family farming which, in turn, are constructed out of ideologies conveying prescriptive messages about the qualities and styles of relationship generated in 'the family'. The agricultural worker, it is argued, is one of the family and, therefore, not exploited (Newby 1977) - as if family labour is free from exploitation. The rhetoric plays extensively on ideologies of local community attachment and supportive interaction between individual families. Thus, 'family', 'community', 'locality', 'farming' and 'rural' all come to mean much the same thing.

These ideas speak of the true virtues and natural justice of England which reside in our countryside - what I have elsewhere (1983) referred to as "village England ideology". Each of these ideological formulae is mediated by the claustrophobic realities of close, pervasive face-to-face contacts at work which characterise the rural LLMs and not just farming, and which provide opportunities for near 'total' class exploitation.

WORK DEPRIVATION AND LOW PAY

Analysis of earnings and wage levels provides the most immediate measures of the degree of exploitation facing particular groups of workers. The distribution of gross weekly earnings for selected occupational groups is set out in Table 4.4 and 4.5. Both pro- and anti-farmer lobbies regularly seize on agricultural wage-rates as some sort of economic talisman against which other rural earnings can be juxtaposed. Therefore, farm workers earnings are used below as a standard of comparison.

The greatest divergence occurs between the senior 'Salaried professionals' (including accountants and university/college teachers) and the rest. Their *salaries* begin where the top 10 per cent of farm-workers *wages* peter out (see Table 4.4). Teachers and nurses ('salaried professionals in health, education and social services'), receiving salaries set by nationally fixed awards (as agricultural workers are supposed to), start earning at around the level of the lowest quartile of agricultural workers, and soon pull further ahead. When the farmworkers have reached their nationally agreed ceiling there are still one-third of these 'lower grade' professionals above them in the earnings table. Clerical and secretarial workers ('non-manuals') hold broad parity with the farm-workers as far as the top quartile of earnings but in the uppermost reaches achieve parity with the highest paid teachers and nurses. The farmworkers also hold parity with the diverse group of workers labelled 'semi-skilled manual workers in transport, construction and servicing'. This group includes warehouseman, roundsman, postal workers and groundsmen. As farmworkers, for our purposes, are coded within the semi-skilled category, this data suggests a remarkable degree of consistency in wage rates received by semi-skilled workers in the rural areas.

Two groups of workers (of the selected grades) do fare worse in the earnings league, at least superficially, when compared with agricultural workers - namely, 'service workers in retailing and allied work' (shopworkers, hairdressers, housekeepers) and 'unskilled manual workers in cleaning and domestic employment'. Both these groups are characteristic of the low-paid sectors where women predominate. The top wages of rural service workers (predominantly 'shopgirls') was about £99 per week in 1981,

Table 4.4 Distribution of gross average weekly income (men and women) (£) for selected occupational grades (all areas)

	Senior salaried Professionals	Salaried professionals in health, education and social services	Non-manual employees	Service workers in retailing and allied work	Agricultural workers	Semi-skilled manual workers in transport, construction and servicing	Unskilled manual workers in cleaning and domestic employment
Lowest decile	122.3	60.5	48.0	26.0	33.7	44.0	26.2
Lowest quartile	180.5	99.0	64.9	33.0	63.0	60.6	48.2
Median	221.2	125.3	75.9	60.5	81.5	85.0	73.5
Higher quartile	296.3	157.6	102.9	67.3	102.9	102.9	84.5
Highest decile	363.2	198.3	184.6	98.9	120.5	120.0	116.9
	N = 22	N = 46	N = 75	N = 26	N = 86	N = 27	N = 24

Source: Author's survey, 1981

Table 4.5   Average gross weekly earnings for various
             categories of employment status (relative to
             GB full-time male non-manual worker = 100

Full-time male	GB	£163.1 =	100
non-manual	All areas	£215.2 =	132
	Shropshire	£291.5 =	179
	Northumberland	£159.3 =	98
Full-time male	GB	£121.9 =	75
manual	All areas	£117.2 =	72
	Shropshire	£131.3 =	80
	Northumberland	£102.8 =	63
Full-time female	GB	£ 96.7 =	59
non-manual	All areas	£104.1 =	64
	Suffolk	£119.2 =	73
	Northumberland	£ 83.1 =	51
Full-time female	GB	£ 74.5 =	46
manual	All areas	£ 66.3 =	41
	Suffolk	£ 78.3 =	48
	Essex	£ 48.2 =	30
Part-time female	GB	£ 43.5 =	27
non-manual	All areas	£ 48.7 =	30
	Shropshire	£ 57.1 =	35
	Essex	£ 42.7 =	26
Part-time female	GB	£ 33.0 =	20
manual	All areas	£ 21.7 =	13
	Northumberland	£ 27.5 =	17
	Shropshire	£ 15.7 =	10

Source: Author's survey and New Earnings Survey, 1981

which was £18 below the top wages for unskilled manual workers and £22 below the top wages for farm workers.

Furthermore, whereas agricultural workers can expect relatively stable employment throughout most of their working lives, 'women's work' in the personal services is a paradigmatic example of, what the dual labour market theorists refer to as, 'secondary sector work'. Even so, only the two salaried, professional groups demonstrate any degree of labour market segregation, according to earnings. These findings on earnings bear out one of the conclusions of Abercrombie and Urry (1983) that 'lower middle strata' groups, represented here by non-manual employees, are being proletarianised, whilst 'upper middle class' groups are becoming an independent service class, relatively segregated - in this case by earnings - from other workers.

The most striking inequalities in earnings are shown in Table 4.5. Average earnings for the various employment groups have been calculated according to the standard of full-time male, non-manual workers in Britain (New Earnings Survey 1981). All the rural non-manual groups have higher indices than the equivalent GB groupings, indicating that the more affluent in these categories choose to live in the countryside. Particular localities evidence specific earnings profiles. The consistently low earnings for full-time employment in the Northumberland area is striking. More startling, however, are the contrasts between equiv-alent groups of men and women. The *highest* area average for women falls below the *lowest* area average for men *in each case*. Despite their generally superior human capital, in the form of educational credentials, women consistently earn less than men in the rural LLMs. The median incomes for women, with equivalent qualifications to men, are at least 20 per cent and most frequently 33 per cent below the men's. In general, *median* incomes for women are equivalent to the *bottom 20 per cent* of the range for men. At the other end of the distribution, the women's top incomes (highest decile) - with the exception of women holding degrees - fall below the top quartile for men. Only two explanations are possible for these divergences between the earnings of equally qualified women and men; either women are segregated into secondary rural LLMs; or they are able to enter the same LLMs as men but, once there, are discriminated against.

In those households where both men and women were working they were paired according to occupational class position (on the Goldthorpe and Hope grading scale). In two-thirds of these 284 cases the man held the higher status job. In the majority of cases where the woman had a higher occupational status than the man, both were employed in relatively low status jobs - typically, she in clerical work, and he in skilled or semi-skilled manual work. In general, therefore, although frequently better qualified than men, women were lower paid in lower status jobs. There were *very* few women in top jobs. Moreover, even in

those instances where women could achieve higher *status*
in the rural labour market than their husbands they were,
nevertheless, lower paid. Male segregation, however,
only takes place at the top. Further down the occupational
class structure of the rural areas women and men weave
complex patterns of relationships around one another,
based on the relative possession of fine gradations of
power and status. In the end, however, the men appear to
have the positional advantage of power, either in the
domestic group or the workplace, to discriminate against
the access of women to equal labour market opportunities.

Locals, like women, appear to be structurally disadvantaged throughout the various rural labour markets,
without these divisions amounting to full segregation on
the basis of ethnicity. Thus, with the exception of the
Northumberland coast area (where S-EG1 is under-represented),
there are between two and three times as many non-locals
as locals in the top class grouping. Throughout the rest
of the class distribution, although locals predominate in
the lower three S-EGs and non-locals in groups 2 and 3,
the differences are far less striking.

Finally, using the index of 'deprivation at work'
devised by Townsend, we are able to compare levels of work
deprivation for class, sex and ethnic groupings (Tables 4.6
and 4.7). The story told by these tables accords with
the previous evidence. In general, proportionately more
women and locals suffer greater degrees of deprivation at
work than men and non-locals. The deprivation score rises
as more of the individual facets of work deprivation are
endured. There are eight component variables of this
scale which, in turn, relate to four specific types of
deprivation: the character of the job; job security; the
working conditions; and, minimal or total absence of
occupational welfare (fringe) benefits (see Townsend 1979,
p. 439).

Two-thirds of the women and more than 80 per cent of
the men were under no, or only slight, threat of encountering deprivation at work. Correspondingly, however,
one-third of the women and 16 per cent of the men suffered
a significant degree of deprivation. Again, the sexes
appear to be divided in terms of labour market position:
women run double the risk that men do of facing deprivation
at work. Local workers also suffered proportionately
greater levels of deprivation than non-locals, though
the differentiation is less acute than between the sexes.

The differences are much sharper when work deprivation
is correlated with social class. Table 4.7 indicates
high levels of *segregation* between the work situations
of the lowest two and the upper four socio-economic groups.
Some 45 per cent of S-EG5 and 69 per cent of S-EG6 suffer
substantial deprivation at work or worse. Significantly,
scores for S-EG4 match those for S-EG1, suggesting that
these 'labour aristocracy' groups are able to exert much

Table 4.6 Percentages of employed men and women, and of employed locals and non-locals, experiencing various levels of work deprivation

Total deprivation (score)		Men	Women	Locals	Non-locals
None	(O)	36.4	31.4	30.6	46.3
Slight	(1-2)	47.3	35.2	44.3	35.1
Substantial	(3-4)	14.1	31.2	22.7	16.1
Severe	(5)	1.9	1.2	1.6	2.1
Very severe	(6+)	0.3	1.0	0.8	0.4
	N =	602	398	647	510

Source: Author's survey, 1981

Table 4.7 Percentages of the economically active in different socio-economic groups, experiencing work deprivation

Total deprivation (score)	1.Professionals	2.Employers and managers	3.Intermediate non-manual	4.Skilled manual	5.Semi-skilled manual	6.Un-skilled manual
None (0)	61.0	52.7	39.1	38.8	10.3	4.6
Slight (1-2)	26.0	44.4	35.9	46.3	45.1	26.1
Substantial (3-4)	9.1	2.8	23.4	13.3	40.8	58.5
Severe (5)	3.4	–	1.0	0.8	2.5	9.3
Very severe (6')	0.5	–	0.6	0.8	1.3	1.5
N =	169	243	192	255	233	65

Source: Author's survey, 1981

greater power than other manual workers to achieve, at
least, adequate working conditions.

Despite high levels of severe deprivation affecting
the lowest grouping there is still an overlap between the
top and the bottom, with some 31 per cent of unskilled
manual workers encountering no, or only slight, deprivation
at work. So, although class differences are more pronounced
than either sex or 'ethnic' divisions, labour market
segregation of occupational class positions remains incomplete. It would, however, be grossly misleading not to
emphasise the very high levels of severe deprivation which
unskilled rural manual workers face at work, which indicate
their substantial separation from the experiences of others
in the rural labour process.

CONCLUSIONS

I began by posing the question: why has the rural working
class remained quiescent in contrast to general labour
militancy and urban social movements in Britain? One set
of answers was to be found, I suggested, in an examination
of rural LLMs. Below, I enumerate five features of the
rural LLMs examined which serve to explain deprivation,
marginality and quiescence amongst the rural working class.

1. The balkanisation of rural LLMs results not from the
interplay of separate institutions structuring fragmentation
but from the structure of the LLM divided, as it is, into
a large number of small enterprises. It is the structural
dominance of propertied capital over industrial capital
that defines the petit-bourgeois structure of both rural
firms and rural LLMs. The structure of the labour markets,
in turn, conditions the general quiescence of the rural
working class. At work labouring groups are isolated into
relationships based on particularistic contact which, in
the absence of any other working class associations, leaves
the individual at the mercy of his or her own bargaining
strength. The depressed level of wages for middle and
lower strata occupational groups, which has been demonstrated,
points to the effective use of dominant class power by
rural employers over their workers. Furthermore, the high
levels of class exploitation experienced by workers, linked
to a complex array of rural ideologies, further reduce the
market capacity of these workers.

2. Rural labour is divided into *segregated* labour markets
along two axes of power. The first axis is the market
capacity of the upper middle strata 'service class'. The
labour market for this group is to be found, both spatially
and socially, beyond the immediate confines of the most
localised rural LLMs. Their means of access to relatively
advantaged occupational class positions is through the
use of formal, achievement-related, qualifications. Once
in 'place' this *class* is able to use various exclusionary

practices - the membership of professional associations, for example - to close out any potential competition. Inevitably, the occupational positions of the service class are almost totally separate from the labour market aspirations of all other rural groups. For reasons given above, the value of acquiring educational credentials occupies a low priority in the socialisation of rural youth, a fact which further strengthens the barrier between the service class and the rest. To this extent the service class exist apart from the dynamics of rural LLM segregation; they merely choose to reside in the countryside.

3. The second axis of labour market segregation is the asymmetrical relations of power *between* groups of 'ethnic locals'. A crucial resource is the capacity to make use of local knowledge and contacts and to exploit these to gain preferential access to higher occupational class positions. Local people who are without this form of market power and lacking formal qualifications tend to be in the lowest and - on other criteria, such as pay and work conditions - the most deprived group within the rural LLM. This group of structurally powerless locals is not merely disadvantaged but is actually segregated into a secondary LLM.

    It is important to clarify a couple of points in respect of these two axes of asymmetry and segregation. Firstly, I am not making a distinction between commuters to urban labour markets and the rest. The service class professionals represent only one small fraction of commuters resident in the rural areas. More than a few ethnically local people are commuters to jobs located in central places, but do not have the power resources and market capacity of the service class. Secondly, the access of rural workers to labour market power, based on localistic association, does not represent a cleavage between locals and newcomers. In general, newcomers do not need to rely on such local market power, from which, in any case, they are normally excluded, based as it is, on kinship connections. The distinction made here concerns the differential exclusionary capacities of separate groups of local people, some of whom are able to exercise greater choice and display enhanced mobility through the occupational class structure, within the distinct confines of rural LLMs.

4. In addition to these two axes of labour market segregation we have found substantial evidence of other cleavages dividing the rural working class into segmented labour groupings. Not surprisingly, these divisions are based on class, sex and ethnicity. Generally speaking, it is apparent that women fare worse than men, locals worse than non-locals and lower grade worse than higher grade workers. These divisions are consistent with patterns of labour market segmentation established throughout Britain.

5. My final conclusion relates to the marginality of the rural working class. As I have shown certain groupings

within the rural labour force suffer from particularly
severe deprivation at work. One feature, however, distinguishes the rural working class, as a whole, and that is
its abject lack of representation on political agendas.
Levels of unionisation amongst rural workers are depressingly
low. Furthermore, none of the major political parties have
demonstrated more than a passing interest in the dynamics
of rural labour markets and the problems of rural deprivation, although great attention has been focussed on rural
conservation and agricultural policy. Rural social policy
is a pitifully neglected subject. In the end, it is their
political marginalisation that constitutes the most severe
problem for the rural working class. Who will act to
challenge these relations of power which structure everyday life in our rural areas? Who will act on behalf of
the rural working class? These questions demand urgent
attention if we are to arrest the processes which have
turned rural workers into one of the most invisible groups
of deprived citizens in contemporary Britain.

NOTES

[1] Some degree of confusion often attaches to the notion of
'localness' in rural social research (see the introductory
chapter, and Chapter 10 of this volume for wider theoretical
discussions of the issue). For the purposes of our survey,
'locals' were any individuals one of whose parents had been
'brought up within the geographical locality delineated by
our study areas'. All other persons were deemed to be 'non-
locals'. Though a number of theoretical and methodological
criticisms, not least those relating to a person's own
perceptions of her 'localness', may be raised against this
definition, it served as a relatively simple 'rule of
interpretation'!

REFERENCES

Abercrombie, N. and Urry, J., 1983. *Capital, Labour and
    the Middle Classes,* (London: Allen & Unwin).

Blackburn, R. and Mann, M., 1979. *The Working Class in the
    Labour Market,* (London: Macmillan).

Bluestone, B., 1970. The tripartite economy: labor markets
    and the working poor, *Poverty and Human Resources,*
    July-August.

Bosanquet, N. and Doeringer, P., 1973. Is there a dual
    labour market in Great Britain? *The Economic
    Journal,* 83, 421-435.

Bradley, T., 1979. Agricultural production, collective
    consumption and class in the British countryside,
    paper delivered to the Xth Congress of the
    European Society for Rural Sociology, Cordoba,
    Spain, April.

Bradley, T., 1983. Within the double margin: poverty and deprivation in rural England, paper delivered to the British Sociological Association Annual Conference, Cardiff, Wales, April.

Brown, M. and Winyard, S., 1979. *Low Pay on the Farm,* (London: Low Pay Unit).

Davis, J.E., 1980. Capitalist agricultural development and the exploitation of the propertied labourer, in F.H. Buttel and H. Newby (eds.) *The Rural Sociology of the Advanced Societies: Critical Perspectives,* (London: Croom Helm).

Doeringer, P. and Piore, M., 1971. *Internal Labor Markets and Manpower Analysis,* (Lexington: Heath).

Edwards, R., Reich, N. and Gordon, D., 1975. *Labor Market Segmentation,* (Lexington: Heath).

Fothergill, S. and Gudgin, G., 1982. *Unequal Growth,* (London: Heinemann).

Giles, A.K. and Cowie, W.F.G., 1957. *An Inquiry into Reasons for 'The Drift from the Land',* (Briston: Department of Agricultural Economics, University of Bristol).

Gillespie, A., 1977. *Journey to Work Trends within British Labour Markets,* (London: London School of Economics).

Goldthorpe, J.H. and Hope, K., 1974. *The Social Grading of Occupations,* (Oxford: Clarendon).

Hale, S., 1971. *The Idle Hill,* (London: National Council of Voluntary Organisations).

Halsey, A.H., Heath, A., and Ridge, J.M., 1980. *Origins and Destinations,* (Oxford: Oxford University Press).

Hunter, L.C. and Reid, G.L., 1968. *Urban Worker Mobility,* (Paris: Organisation for Economic Cooperation and Development).

Johnson, T., 1977. The professions in the class structure, in R. Scase (ed.) *Industrial Society: Class, Cleavage and Control,* (London: Allen & Unwin).

Kerr, C., 1954. The balkanisation of labour markets, in E.W. Bakke (ed.) *Labor Mobility and Economic Opportunity,* (New York: Wiley).

Kreckel, R., 1980. Unequal opportunity structure and labour market segmentation, *Sociology,* 14, 525-550.

Larson, M.S., 1977. *The Rise of Professionalism: A Sociological Analysis,* (Berkeley: University of California Press).

Levitan, S.A. *et al* 1976. *Human Resources and Labor Markets,* 2nd edition, (New York: Harper & Row).

McDermott, K. and Dench, S., 1983. *Youth Opportunities in a Rural Area: A Study of Mid-Wales,* (Sheffield: Manpower Services Commission Research and Development Series).

Metcalf, D. and Richardson, G.R., 1972. Labour, in A.R. Prest and D.J. Coppock (eds.) *A UK Manual of Applied Statistics,* (London: Weindenfeld).

Minchinton, W. (ed.), 1968. *Essays in Agrarian History,* (Newton Abbot: David & Charles).

Moseley, M.J., 1980. *Rural Development and its Relevance to the Inner City Debate,* (London: Social Science Research Council, Inner Cities Paper 9).

Nalson, J., 1968. *Mobility of Farm Families,* (Manchester: Manchester University Press).

Newby, H., 1977. Paternalism and capitalism, in R. Scase (ed.) *Industrial Society: Class, Cleavage and Control,* (London: Allen & Unwin).

Newby, H., 1977. *The Deferential Worker,* (Harmondsworth: Penguin).

Newby, H., 1980. *Green and Pleasant Land?* (Harmondsworth: Penguin).

Offe, C. and Hinrichs, K., 1977. *Socialokonomie des Arbeitmarktes und die Lage "benachteiligter" Gruppen von Arbeitnehmern,* (Luchterhand: Neuwied-Darmstadt).

Robinson, D., 1967. Myths of the local labour market, *Personnel,* December.

Shaw, M. (ed.) 1979. *Rural Deprivation and Planning,* (Norwich: Geo Books).

Shucksmith, M., 1981. *No Homes for Locals,* (Farnborough: Gower).

Smart, M., 1974. Labour market areas: uses and definitions, in D. Diamond and B.P. McLaughlin (eds.) *Progress in Planning,* 2(4), (Oxford: Pergamon).

Townsend, A.R., 1981. Geographical perspectives on major job losses in the UK 1977-80, *Area,* 13(1).

Townsend, P., 1979. *Poverty in the United Kingdom,* (Harmondsworth: Penguin).

Wenger, C., 1980. *Mid-Wales: Deprivation or Development,* (Cardiff: University of Wales Press).

Winter, H., 1980. *Homes for Locals?* (Exeter: Rural Community Council of Devon).

Winyard, S., 1982. *Cold Comfort Farm,* (London: Low Pay Unit).

5. The rural labour process: a case study

of a Cornish town

J. HERMAN GILLIGAN

A number of writers have emphasised the significance
of the sub-regional/local level in the analysis of the
restructuring of wider spatial divisions of labour (Fothergill
and Gudgin 1982; Massey 1978, 1979; Urry 1981; Chapter Three, this
volume). This chapter presents such an analysis with respect to
the small Cornish coastal town of Padstow. Although largely
subsumed under the amorphous 'South-West Region'. Cornwall
has recently become a focus of interest for policy-related
research (McNabb and Woodward 1979; Crine and Playford 1981)
and regional and industrial geography (Massey 1983; Shaw
and Williams 1981; 1982; Perry 1983). Within empirical
sociology, however, Cornwall remains neglected.

The British 'community studies' of the 1950s and 1960s,
biased though they were towards the highland zone and the
Centic fringe, did not produce a Cornish monograph (Symes
1981, p. 63); and despite the contemporary resurgence of
interest in the anthropology of rural Britain (Cohen 1982:
Ennew 1980; section three of this volume), no ethnographic
study set within Cornwall has yet been published.[1] Whether
this omission will be remedied by a "linked series of
community studies", established to "provide a geographically
more representative portrayal of the impact of recession",
as recommended in a recent report (Newby 1982, p. 69),
remains to be seen. This chapter, meanwhile, seeks in a
modest way to counteract such neglect. It presents first
a brief account of the social and economic context of
Cornwall, before proceeding to an historical case study of
labour and employment in Padstow.[2]

THE CORNISH CONTEXT

Cornwall is a geographically remote peninsular, greatly
subject to the influence of the sea. It has a Celtic
history, linguistically and culturally distinct from England.
Hechter's dismissive assertion that "the Celtic region of
Cornwall became largely assimilated to English culture by
the mid-seventeenth century" (Hechter 1975, p. 64), has

been challenged in brief (Rallings and Lee 1978), although
no detailed study of the process of 'anglicization' has
been undertaken.

Hechter makes reference to Cornwall's substantial trade,
and its integration into the English economy prior to 1600.
As early as the fifteenth century, Cornwall had an exceptionally diversified economy, which, in addition to a very
prosperous mixed agriculture, included mining, fishing,
shipping, textiles, quarrying and shipbuilding. But it was
the capitalist development of tin and copper mining during
the eighteenth century, largely concentrated in the centre
and west of the county, which provided the basis for industrial growth up to the 1850s. Its subsequent economic
history has been characterised by severe fluctuations
in the predominant primary extractive industries (as well
as in agriculture and fishing), leaving Cornwall as "Britain's
first decayed industrial area" (Rallings and Lee 1978, p. 7).
The decline in mining led to successive waves of emigration
during the second half of the nineteenth century, resulting
in a continuous population fall from 369 000 in 1861 to
322 000 in 1901, in sharp contrast to national trends.

It has been suggested that the scale of emigration by
miners, particularly during the period of expanding unionisation in other parts of the country, significantly depleted
Cornwall's industrial proletariat and effective trade union
base, and accounted in large part for the failure of the
emergent Labour Party to penetrate the county (Rallings
and Lee 1978). Other characteristics of Cornish political
culture are a strong Liberal/Non-conformist tradition, and
a nationalist consciousness originating from a late-nine-
teenth century linguistic and cultural revival, manifest
politically in the Mebyon Kernow (Sons of Cornwall)
movement and the Cornish Nationalist Party.

Perhaps the single, most significant event in Cornwall's
economic history since the heyday of tin and copper mining,
was the opening of the Saltash bridge in 1859, which
provided a railway link between Cornwall and England across
the River Tamar. "It is not too much to say that this
wrought, in succeeding years, a revolution in the county's
economy" (Simmons 1959, p. 13). Early capital investment
in railway schemes in Cornwall sought to exploit the
possibilities of greatly expanded markets for agricultural
produce and fish, and well before the turn of the century
the railway had also become the harbinger of tourism.
Railway company amalgamations and extensive publicity
campaigns stimulated the development of the tourist industry,
which has progressively overshadowed the social and economic
life of Cornwall.

The industrial structure of modern Cornwall reflects
the overlapping configuration of the historical forces
sketched above. The primary and service sectors make a
disproportionately large contribution to both male and
female employment, whereas the manufacturing sector remains

small-scale and provides relatively few jobs. Agriculture
and fishing are still significant for male employment, as
is mining and quarrying - almost exclusively for china
clay. Unlike tin or copper production, the china clay
industry encountered little foreign competition and has
continued to expand to become the county's largest single,
industrial employer. Service industries, in particular
those relating to tourism, employ the highest percentage
of the female workforce (Crine and Playford, 1981).

In April 1982 Cornwall was the county with the lowest
average gross weekly earnings for men in full-time employment, £28 lower than the figure for the whole of Great
Britain. For women in full-time employment, Cornwall
had the eighth lowest average earnings of any county,
almost £10 lower than the national average (Department of
Employment 1983b). As has been noted,

> The effects of low regional earnings
> and an employment structure biased
> towards low paying industries inter-
> relate very strongly. In effect, the
> problem of low earnings in Cornwall
> is an exceptional reliance on low
> paying industries which pay especially
> low wages in this part of the country.
> (Cornwall County Council 1976, p. 39).

For many years, the unemployment rate in Cornwall has
been consistently higher than elsewhere. Between 1956
and 1975, for example, the average annual rate for the
county was 4.8 per cent, while that for Great Britain as
a whole was 2.3 per cent (Cornwall County Council 1976,
p. 107). This pattern has persisted throughout the current
recession: the unemployment rate in Cornwall of 9th December
1982 and 9th June 1983 was 17.5 per cent and 13.7 per cent
respectively, compared to a national average of 13.1 per
cent and 12.3 per cent (Department of Employment 1983a).
Seasonality in unemployment is very marked in Cornwall,
especially in tourist-related activities, where both men
and women are affected, but also in the male-dominated
fishing, construction and agricultural industries. It
has been estimated that male unemployment in Cornwall in
winter is one and a half times that in summer, and for
females, two and a half times the summer figure (Cornwall
County Council 1976, p. 109).

After a century of stagnation and decline, the county's
population has recently begun to increase sharply, in
large part the result of in-migration predominantly by
people over retirement age. The inter-censal increase
between 1971 and 1981 was 13.25 per cent, compared to
0.83 per cent for England and Wales as a whole; and by
1981, 21.7 per cent of the county's population were of
pensionable age.

PADSTOW'S ECONOMIC HISTORY

Padstow is situated on the north Cornish coast, 15 miles north-west of Bodmin and 15 miles north-east of Newquay, the county's premier resort. The town of 2200 inhabitants shelters in a valley on the western side of the broad Camel estuary, protected from the open sea by a striking promontory. As depicted on postcards, it presents itself as the stereotypical Cornish fishing town, an attractive amalgam of old slate and stone houses huddled around a focal harbour, enjoying a panoramic view of headlands and estuary. It is a compact settlement of early origin, accessible by road only from the south and the west, and remains the service centre for the immediate agricultural hinterland and the coastal satellite villages and hamlets. The market town of Wadebridge, eight miles eastward, provides the nearest upstream bridging point across the river Camel. Padstow is the only sizeable inlet and harbour along the inhospitable north coast of Cornwall, and given this strategic position, the historical signif-icance of the town has been related to its associations with the sea.

Carew, in his Survey of Cornwall in 1602, described Padstow as "a town and haven of suitable quality, for both (though bad) are the best that the north Cornish coast possesseth" (Carew 1811, p. 340). It had emerged as a thriving port during the medieval period, its first stone pier being built before 1536. The subsequent diversification of trading activity has been indicated by Pearse (1963, p. 133):

> The beginning of the expansion of
> mining in Cornwall was already
> reflected in Padstow's shipments
> of copper ore to Bristol between
> 1690 and 1700. Refined tin, a long
> established article of export,
> continued to be shipped. Later still,
> in the eighteenth century, we find
> antimony and lead ores amongst
> Padstow's exports. Cheese, wheat,
> barley, oats, cured fish of many
> types, and slate stones were
> regularly sent away.

> Padstow imported timber for
> every type of use, woollen cloth,
> glass, salted hogs and tallow from
> Ireland; coal from South Wales; salt
> from France, and later from Liverpool;
> linen and canvas from Brittany; peas,
> malt and hardwares from the Severn
> estuary ports; a wide range of products
> from France and, in the nineteenth
> century, timber from Norway and Sweden,
> and hemp, iron and jute from Russia.

By the mid-nineteenth century, at the height of its
commercial prosperity, Padstow had become a port of emigration,
a regional centre of national and international trade, and
an established shipbuilding town.

Few improvements to the port were undertaken until
the nineteenth century. In 1827, under the auspices of
the Royal National Lifeboat Institution, Padstow received
its first lifeboat. Two years later the Padstow Harbour
Association was formed, and was instrumental in setting up
a branch of the Trinity House Pilots, and erecting three
capstans and hawsers at Stepper Point, where the estuary
meets the sea. But the most economically significant
stimulus came in 1844 when the Padstow Harbour Act "for
regulating, maintaining and improving the port of Padstow"
vested the management of the port and its facilities in
an elective body of Harbour Commissioners. The Lord of
the Manor of Padstow, whose Elizabethan family seat overlooks the town, conveyed the quays and the right to tolls
to the Commissioners, who in large part represented the
leading merchants and shipowners of the locality.

However, trading and shipbuilding prosperity was short-
lived. The success of Padstow's shipyards was based on
the production of sleek wooden sailing schooners, but with
the advent of iron ships and steamers, rapid decline set
in. Sea-borne trade also suffered with the general
reduction in coastal traffic as a consequence of the
development of an embryonic national railway network.
Heavy unemployment resulted, and the parish population
declined from a peak of 2489 in 1861 to 1877 in 1891. Many
shipbuilding workers, a number of whom were migrants to
Padstow seeking regular work, left for the dockyards of
Plymouth, Portsmouth, Cardiff and Swansea, or emigrated
to America. By 1900 all that remained of a once thriving
industry was a handful of self-employed boat repairers and
one boatyard (Table 5.1).

The severest long-term effects of the decline in
shipbuilding on the local economy were mitigated by a
revival in the fisheries after the turn of the century,
as a result of the extension of the railway to Padstow.
Ambitious early railway plans for the area did not come to
fruition: the Cornwall Railway Company obtained Parliamentary powers to provide a line to Padstow in 1846, but this
scheme was abandoned six years later. After much local
agitation the North Cornwall Railway Company, having made
a working agreement with the London and South Western,
received an Act in 1882 to provide a line for the 50 miles
from Halwill to a terminus by the dockside at Padstow, but
due to financial stringency the final section was not
completed until 1899 (Williams, 1973).

With the arrival of the railway, fish could be landed
in quantity and transported direct to the lucrative London
markets. At that time, however, few Padstow people
operated commercial fishing boats, and the overwhelming

95

Table 5.1   Shipbuilding in Padstow

Date	Number of ships built	Tonnage
1800–1809	8	393
1810–1819	14	745
1820–1829	27	1559
1830–1839	37	2190
1840–1849	41	2058
1850–1859	55	2783
1860–1869	62	6275
1870–1879	36	4361+
1880–1889	8	989
1890–1899	2	19+
	290	21372+

Source:

These figures have been aggregated from a comprehensive
list compiled from primary sources, and kindly supplied
by Mr G Farr of Bristol. They refer only to ships
built at sites within Padstow harbour, and not to other
shipbuilding sites within the Padstow Port of Registry.
The tonnage totals for 1870–1879 and 1890–1899 are
incomplete as in certain cases individual ship tonnages
are omitted from the original records.

bulk of port traffic originated elsewhere. Before the
First World War, Brixham sailing trawlers from Devon and
Scots herring drifters were regular port users. An
indicator of the scale of such activity is contained in
the Census return for 1911, where the inflated town population of 2480 included the enumeration of 773 visiting
fishermen aboard 154 vessels.

In the 1920s abundant herring shoals off the north
coast of Cornwall were fished by small boats from other
Cornish ports who landed their catches at Padstow. But
the most significant seasonal presence during the inter-
war period was that of trawlers belonging to the east
coast company-owned fleets from Lowestoft and Yarmouth.[3]
The speculative use made of Padstow by non-local boats is
indicated by Table 5.2, which compares Padstow's rank in
the national league table of ports with that of Newlyn,
the major commercial fishing port in Cornwall, which had an
established local fleet. The evident decline in Padstow's
importance as a fishing station was due to two factors:
firstly, by the 1930s herring had largely disappeared from
the coast in any quantity, as a result of overfishing and
shoal migration; and secondly, technological developments
within the trawling industry meant that the east coast
fleets no longer found Padstow a viable landing port. As
a result, by the 1950s fishing had declined just as shipbuilding had done.

At the outbreak of the Second World War, three Royal
Air Force and Fleet Air Arm stations and airfields were
created a few miles to the south of Padstow. After the war,
the 'Raf' presence became established when 105 houses were
built on the outskirts of the town to accommodate forces'
personnel and their families. An extensive local authority
building programme also began in the late 1940s and 1950s,
to rehouse the inhabitants of over one hundred sub-standard
and overcrowded, pre-1880 cottage dwellings in 'downtown',
the old town centre (Cornwall County Council 1952, p. 154).
During this period, building firms provided much needed
employment, albeit mainly temporary or casual work. In
addition, the china clay industry, based some 25 miles away
around St Austell, recruited construction labour for pipe-
laying and tank-building gangs, and came to be regarded as
an employment safety net for those who would entertain
travelling such a long distance to work. The disappearance
of fishing as a major source of employment was therefore
compensated for; but in turn, the significance of these
industries became overlaid during the 1960s with the
advent of mass tourism.

The railway which had provided such a stimulus to the
fisheries also brought the tourists. Ironically, nineteenth
century Padstow, with its industries and national and
international trading links, was a cosmopolitan centre well
accustomed to transient residents. But these visitors
were on business, not seeking leisure. Indeed, early
travel literature was far from flattering about a town with

Table 5.2   National Sea Fisheries Ranking
             of Ports by Volume of fish landed

Date	Padstow rank	Newlyn rank	Total no. of ports listed
1922	19	14	171
1925	19	12	175
1930	17	15	179
1935	18	14	169
1937	15	14	169
1940	–	–	–
1946	83	14	156
1950	28	10	151
1953	130	–	140
1955	122	8	142
1960	97	7	142

Source:

Compiled from various annual Sea Fisheries
Statistical Tables published by the (then)
Ministry of Agriculture and Fisheries.

a working harbour as a potential tourist attraction. One
handbook describes Padstow as "one of those antiquated
unsavoury fishing towns which are viewed most agreeably
from a distance" (Murray 1865, p. 223); and a later guide-
book says, "we imagine that no one deliberately visits
Padstow for its own sake" (quoted in Page 1897, p. 126).
However, as early as 1880 a local advertising committee
expounded the virtues of Padstow's climate, bathing, boating,
fishing and coastal scenery in the *Western Morning News,*
to "tourists, artists, invalids, and others who desire to
spend their summer holidays pleasantly".

Between 1880 and 1907 new building was almost static
(Cornwall County Council 1952, p. 154). The exception was
the vanguard of the new growth industry, an opulent fifty-
bed hotel completed in 1900, adjacent to the railway station
and physically dominating the town. The early development
of tourism catered largely for upper middle class visitors,
who were accommodated in hotels in the town, or in the
burgeoning number of coastal villas built as second homes
or rented out for annual holidays. However, unlike the
nearby resorts of Bude and Newquay, mass tourism did not
appear in Padstow until the 1960s. The tourist trade was
exploited by external capital, much of it small capital
and family enterprises. The influx of these entrepreneurs
as residents, combined with the increasing number of
retired in-migrants and second-home owners, created a
social and cultural cleavage between the native Padstonians
('Padsta people') and the outsiders ('strangers', 'furriners',
'up-country people'). Padstonians resented the 'taking
over' of 'their' town, but saw themselves as economically
and politically powerless, progressively becoming an
encapsulated community, residentially outcast to the
the council housing estates, as outsiders gentrified older
property and bought up shops and businesses.

THE STRATEGIES OF LOCAL LABOUR

The admixture of types of economic activity characteristic
of Padstow in the first half of this century resulted in
a specific labour market experience: a combination of
synchronic bricolage and diachronic occupational transition.
This experience is captured in the following occupational
history of a male Padstonian born in 1901.[4]

> When I left school I went to work with
> me father as a shipwright. He used to
> work on his own, repairs and stuff. I
> was there for a few years with him, but
> work was getting slack. So I went down
> to Strang's (boatyard) for a while, but
> again work went slack and I had to go on
> the dole.
>
> Well, I was friendly with Bill's
> father, and he had a fishing boat. He

said the old chap that was with him wasn't
very good, and why don't I come on instead.
Well, I'd never done much like that,
only been in punts (small rowing boats)
before, but then I stayed with him.

'Course, the herring catching was
good then. The herring season off Padstow
was the end of October, November, and
say up to the third week in December.
But that was about it. Well then they
(the herring) used to move on, round
Plymouth and Brixham. You'd get round
there just before Christmas, that would
be to February, and then we'd come back
and work on the trawlers till about the
middle of May, and then we used to start
trawling or crabbing (in our own boat).
And we used to wait here until perhaps
July and then we would go to Bideford to
work for a time till we could get back
for the herring round here.

The herring season was when everybody in town had got money. Well the
shopkeepers used to give them what they
wanted until the herring come. As soon
as the herring come, everybody used to
square up then. But until then it was
tight. What finished it though was people
buying boats and trying to go fishing,
or trusting someone else to run a little
boat for them. Well, a lot lost their
money, 'cause the herring went.

So we were fishing most of the year
except when the trawlers come, February
time. With the east coast trawlers we
used to be packing (fish), working on
the (fish) market, and in the fish
stores. There was work for all. I
mean each trawler had, what, a hundred
ton of coal to put aboard in a day. And
they all had to be supplied, with ice
and all. You could get a job putting
coal, or anything like that. But the
Lowestoft boys brought their own people
(trawler crews), so the only thing was
working ashore. And they had their own
officials, their own auctioneers, and
men in charge of the boats. But there
was plenty of work, like the rough work.

After I finished fishing, just before
I got married, I packed up on the water
and got a job. I worked on building the
dock wall (a government-funded employment

project in 1933). I was looking after
the barge down there with the crane and
all on. Well, the firm that built that,
he built some houses up Tremewan Road,
and I went up there for a bit. And
the boss he had a motor launch here, and
he asked me if I'd look after that. Well,
it was only there a couple of years, and
he went to Newquay with it. He wanted
me to go down there, but I said no, I
wasn't shifting to go down there. Well
then I finished with them. I did a
little bit for Penmain's (agricultural
and supplies merchant). That was casual,
it all depended on what boats come in.
Then I went on the council (worked for
the local authority as a general labourer).
And when the war started I got permission
to change, to this Jago's job (builders'
merchant). I was a foreman down the yard
there, and I stayed till I retired
(in 1966). We had about five working
down there. And we used to employ a lot
of casual, most of it was casual
work, putting timber, cement, coal, say for
a couple of days, and if it was quiet,
they'd go some other place, on the farm
or something like that.

And of course I'd been with the
Lifeboat for years. My first trip was
December 1921. When I packed it up, I
went on the slip, to bring 'em in.
And I carried on after that to call 'em
out. That's how I got the phone. They
put it in for me. Well, I did it
'cause I was down on the quay all the
day. But that was a tie, twenty four
hours a day you was on call.

While there was no 'occupational community' deriving
from the dominance of one particular industry (despite the
prevalence of fishing), there was a commonality of experience in the annual round of seasonal, temporary and casual
labour; self-employment, employment and unemployment;
skilled, semi- and unskilled labour, within a range of
allied occupational spheres and familiar work environments.

Padstonian working class men of this, and subsequent,
generations were socialised from an early age into the
range of skills associated with work 'around the quay'.
Familiarity with fishing methods, marine navigation, and
boat and engine maintenance, was typically well developed
by school-leaving age, and woodworking and construction
experience was often gained in youth, as well as that of
agricultural and dockside work. The development of such a
repertoire of skills was the primary survival strategy for

men. For women, the extension of domestic labour offered
opportunities for paid employment in hotels, boarding
houses or pubs, and in domestic service.[5]

Padstow's economic history, the continued lack of
secure full-time work, depressed wage levels, and the
reliance tody on tourism, construction and fishing (all
largely seasonal, high-risk industries), conspire to ensure
the preservation of established and extensive flexible
participation in a combination of formal and informal
economic activities. This is not to romanticise the
familial and communal support networks which have been
developed by Padstonians in the face of adversity, or to
imply that it has ever been possible for many to obtain
anything approaching a comfortable living by engaging in
informal economic activities. The informal economy does not
comprise an alternative economic system, for it is based,
like the formal economy, on the ownership of property
(Henry, 1982). However, such a popular tourist area does
provide enhanced possibilities of income generation for
owners of even modest capital resources.

A contemporary example of a complex combination of
economic activities is provided by three inter-related
households of the Penrose family:

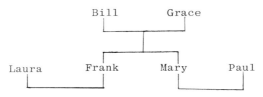

Paul is a Padstonian in his thirties
who left school at 15 to join the Army.
After completing nine years' service, he
returned to Padstow to live with his
parents. He has since been casually
employed as a general building labourer
for a number of years, working on various
sites in north Cornwall, as well as
further afield. His wife Mary, also a
Padstonian, worked for the local authority
as a typist before they married.

They live in a large, four-bedroomed
Victorian house on 'The Gold Coast' (the
more expensive private housing area).
The house was originally bought by Mary's
paternal grandfather, who was a shipyard
owner, and Mary's parents lived there themselves until Paul and Mary were married in
1976. The property is maintained as a
guest house in the summer, being able to
accommodate up to seven visitors. At the

102

1980 bed and breakfast rate of £4.50 per night, it provides a valuable source of income for three or four months of the year. During the winter they have also taken in lodgers.

Mary does most of the work involved in the day-to-day running of this venture, although at the height of the season she employs a young housewife who lives nearby to launder and clean for two hours a day during the week. When particularly busy, Mary's mother also 'helps out'. Mary's parents, Bill and Grace, left the large house and moved to 'downtown', into a small, two-bedroomed cottage which they inherited on the death of one of Grace's aunts. For many years Grace also used to cater for visitors in the large house, but as she is now in her sixties, she is quite content to leave such labour-intensive activity to her daughter. In 1974, just after Bill retired from his job with the Water Board, they bought another old, near-derelict shell of a cottage in Padstow. Grace now deals only with the bookings for this cottage, which has been extensively modernised by Bill and their son-in-law Paul, and is let out to self-catering holidaymakers at a peak season rate of £60 per week.

Bill and Grace's son, Frank (Mary's elder brother), lives with his wife Laura and their two teenage sons in a three bedroomed house in the next street to his parent's cottage. Frank is employed as the head gardener for a small market gardening firm, where his eldest son also works.

When Laura's children were younger she also used to take in summer visitors. For a number of years the boys were accustomed to either sharing a bedroom or sleeping in the same room as their parents in order that Laura could cater for up to four paying guests. But now that the mortgage has been paid off, and their eldest son is working, there is no longer the economic imperative to gain additional income by such means, which inevitably encroached upon the family's domestic life.

However, Frank and Laura currently have another source of supplementary

income. With the aid of a bank loan,
Frank bought a small 18 foot boat from a
retired local fisherman. Being in need
of repair, the boat was housed in a
friend's store over the winter while Frank
and his father restored it in preparation
for the coming tourist season. In the
summer it carries up to eight paying
passengers a time on hour-long mackerel
fishing trips. Although owned by Frank,
the boat is operated by Paul, his brother-
in-law, who in return receives a share of
the accrued profits.

THE CONTEMPORARY EMPLOYMENT PATTERN

Tables 5.3 and 5.4 outline the contemporary employment
pattern of Padstow. Table 5.3 indicates the extent of self-
employment (totalling 26.6 per cent of economically active
men), connected with the construction and fishing industries,
and the service sector. The high female activity rate
(71.3 per cent of women) is due very largely to the significance of tourism, which accounts for much of the available
part-time and seasonal work, and stimulates further
employment for women in other service industries.

Table 5.4 highlights the preponderance of the service
sector for full-time female employment (50.9 per cent are
employed in retail distribution, and hotels and catering).
For men, construction is the major full-time industry,
stimulated also by the tourist trade. Fishing, as a main
occupation, engages a larger number of men than hitherto,
either as crewmen on company-owned trawlers fishing out
of Newlyn, or as Padstow-based fisherman-owners, combining
'crabbing' for shellfish in the summer months with long-
lining and/or light trawling.

The data above relates to responses elicited to questions
asked by the researcher regarding paid employment. As
such, it is interesting to compare the percentages in
Table 5.3 for persons presenting themselves as 'out of
employment' (5.3 per cent for males and 0.5 per cent for
females), with the average annual unemployment rates over
the same period registered by the nearest Employment Office
in Wadebridge (see Table 5.5).

A relatively small proportion of those employed are
involved in fully capitalist wage relations. The full-
time male extractive workers represent the largest number
of persons employed in a single industrial workplace: at
the English China Clays pits in the St Austell area.
Manufacturing industry, other than a printing works, is
located in small factory premises on the Wadebridge and
Bodmin industrial estates. With the exception of the town's
largest hotel, which is owned by a prestigious national chain,
service employment in Padstow is on a very small scale.

104

Table 5.3   Padstow 50% sample household survey:
             Employment status

		Men Aged 16-64		Women Aged 16-59	
		(No.)	(%)	(No.)	(%)
	(employed (	110	48.5	43	22.1
	(self-employed, (no employees (	35	15.4	3	1.5
Economically active	(self-employed, (with employers (	16	7.0	7	3.6
	(part-time or (seasonal work only (	19	8.4	85	43.6
	(out of employment	12	5.3	1	0.5
	(permanently sick (or disabled (	7	3.1	1	0.5
Economically inactive	(full-time student (	11	4.8	5	2.6
	(others including (wholly retired (and housewives	13	5.7	47	24.1
	unclassified	4	1.8	3	1.5
		227	100.0	195	100.0

Source:

This, and the following table, relate to data obtained from
a 50% sample household survey of Padstow (excluding the
RAF married quarters' estate), which was carried out over
the fieldwork period.

Table 5.4   Padstow 50% sample household survey:
            Full-time employment by industry group

	Men Aged 16-64		Women Aged 16-59	
	(No.)	(%)	(No.)	(%)
Agriculture and horticulture	5	2.2	1	0.5
Fishing	17	7.5	0	
Extraction of minerals	13	5.7	0	
Metal goods, engineering and vehicle industries and other manufacturing	14	6.2	2	1.0
Printing and publishing	4	1.8	5	2.6
Construction	32	14.1	0	
Retail distribution	17	7.5	15	7.7
Hotels and catering	11	4.8	12	6.2
Transport services	7	3.1	1	0.5
Postal services and tele-communications	6	2.6	1	0.5
Banking, finance and business services	3	1.3	2	1.0
Public administration	11	4.8	2	1.0
Education	6	2.6	7	3.6
Other services	10	4.4	5	2.6
Armed forces	3	1.3	0	
Other	3	1.3	0	
	162	71.2	53	27.2

Table 5.5   Average Annual Unemployment: Wadebridge
             Employment Office

	Male	Female	Total
1979	12.03%	7.3%	9.98%
1980	13.65%	7.5%	10.98%

Source:

Table constructed from quarterly unemployment
figures for the Wadebridge Employment Office,
supplied by the Manpower Services Commission
South West Regional Office, Bristol.

THE PADSTOW INDUSTRIAL ESTATE

Local politics during the 1960s was dominated by the
'light industry' question. The conflict of interests
among residents, between Padstonians and outsiders, crystallised around the issue of encouraging light industry
as against tourist development in order to boost regular
employment in the town. Views were emotively expressed in
early editions of the *Padstow Echo* magazine, founded by a
Padstonian in 1964:

> A big majority of the people of
> Padstow are in favour of a light industry
> in the environs of the town ... we can
> only assume that a certain minority of
> persons would suffer if an industry offered
> all-the-year-round congenial work to young
> people and married ladies who had, until
> then, been readily available for part-time
> seasonal work in hotels, cafes, etc.
> (*Padstow Echo,* 10 March 1965, p. 17)

> Padstow had a staple industry namely
> shipbuilding which was not limited to four
> summer months. There was little unemployment
> and Padstow's community was well balanced in
> its age groups ... To-day the position is
> changing fast - there is an exodus of young
> people from the town because of a lack of
> light industry ... If action is not taken
> quickly the future of Padstow will be
> limited to catering for retired people and
> property speculators. It is up to all
> Padstonians to support any attempt to bring
> light industry to the town.
> (*Padstow Echo,* 10 March 1965, p. 34)

Padstonians saw industrial development as providing the key
to more stable employment opportunities. To tourism
entrepreneurs, such development was regarded as a threat to
the town's further tourist potential, and would of course
severely deplete their reserve army of seasonal labour.
The retired in-migrants viewed the possible expansion of
either activity with suspicion, fearing the despoilation of
the environmental haven they were anxious to preserve.

Two possible sites for an industrial estate were considered by the Rural District Council. One was on agricultural land to the south-west of the town, and the other
was the redundant dockside railway land used only for car
parking after the closure of the line to Padstow in 1967.
The latter would have provided a prime site for marine
engineering development or other marine-related activity
with which Padstow had been historically associated.
However, it was the inland site which was eventually purchased
and developed. While the introduction of new industry
became a source of open conflict, the traditional activities
of shipbuilding and fishing, as seen from all three perspectives, assumed an idealised character. To Padstonians
they symbolised part of their 'heritage'; to tourism entrepreneurs they became culturally commoditised as 'local
colour'; and to the in-migrants they represented intrinsic
elements of their 'village of the mind' idyll of a Cornish
community.

The opposition between 'continuity' and 'change' has
provided the framework of many community studies. In
Padstow, 'continuity' can be seen as being represented by
the maintenance of the close-knit, numerically significant,
and self-conscious, local Padstonian population. Perhaps
tourism is therefore to be seen as the agent of 'change'?
But Padstow's economic history is replete with examples of
the accommodation (in both senses of the term) of transient
or temporary residents for varying periods of time; and the
development of the tourist trade has also maintained the
type of labour market experience previously encountered.

The exploitation of tourism has sharpened the Padstonian/
outsider distinction and exposed conflicting interests, but
the perceived threat has served to enhance the cultural
significance of what it means to be 'local', and engendered
an almost exaggerated sense of Padstonian communal solidarity. However, this sense of solidarity is presently in
danger of being undermined by the successful development
of the very intervention which was sought by Padstonians
in order to help maintain the local community. The
industrial estate, which in its first years of existence
was little more than the relocation site for a number of
small local businesses, received a significant injection
when a London-based printing company moved there in 1973.
In 1980 the firm employed about 40 full-time staff, with
a planned extension of the premises to increase the labour
force by another 25 to 30. It offered regular, all-year-
round, secure employment for both men and women in a modern

factory - with relatively high wages, pleasant working
conditions, and the possibility of training. Most of the
skilled employment was taken up by experienced printing
workers recruited nationally through their trade unions,
but semi-skilled printing and binding jobs were made
available to locals, as was other office, canteen, ware-
housing and cleaning work.

For those who were able to obtain employment at 'The
Press', fixed shifts were the norm, with overtime working
when schedules had to be met. The maximisation of the
income opportunities presented, required a degree of reliance
on familial or communal support for domestic and child-
caring services as to threaten established norms of
reciprocity. The system of generalised exchange upon which
working class households had depended in order to optimise
existing resources, was predicated on notional equal access
to available employment. This access was obtained via the
informal networks within which Padstonians were located.
The presence of 'The Press' disturbed that equilibrium,
by unevenly distributing highly sought-after income and
employment, and provoking a degree of conflict and bitter-
ness amongst unsuccessful applicants. Gender relations
were also affected, as wives now had the possibility of
earning more than many full-time employed husbands. And
the unionisation of the workforce nurtured a trade union
consciousness, particularly amongst the female employees,
which was quite novel for Padstow.

Since the completion of my fieldwork in mid-1980, the
ownership of 'The Press' has changed hands, and industrial
relations would appear to have become less amicable. But
perhaps this new industry is only to be seen as just another
layer of the palimpsest that is Padstow's economic history,
to be overlaid by ...?

NOTES

[1] However, I am aware that two social anthropologists currently have work in progress. Michael Ireland, of the Department of Sociology and Anthropology at the University College of Swansea, is working on a case study of tourism in Cornwall, based on fieldwork in Sennen, near Land's End. Connie Brown, of the Department of Sociology at the University of Manchester, is investigating community identity and its manifestation in everyday life, within the parish of Zennor, also in south-west Cornwall.

[2] I do not wish to imply that Padstow can be seen as a 'microcosm' of Cornwall. Rather, I seek to draw attention to broader trends as they are manifested at the local level, without detracting from the specificity of the local context. This chapter is based upon doctoral research carried out between 1978 and 1981, the main fieldwork period being from March 1979 to August 1980. Padstow was selected quite fortuitously as a fieldwork site, and not for any reason of supposed 'typicality'. For an account of the research process, see Gilligan (forthcoming).

[3] For a detailed account of the variety of socio-economic forms of fishing activity, see Thompson (1983), the first major sociological study of fishing and fishing communities in Britain since Tunstall (1962).

[4] Names and certain details have been changed in the ethnographic data presented in this chapter.

[5] Thompson (1983) emphasises the direct contribution of women's labour to fishing. In Padstow, female employment was largely related to just one specific task, carried out by non-locals: during the herring season, a peripatetic group of Scottish women (known as the 'kipper girls') were employed to gut and smoke the fish in dockside sheds, before it was sold and transported to London.

REFERENCES

Carew, 1811. *Survey of Cornwall* (London) (first published 1602).

Cohen, A.P. (ed.), 1982. *Belonging* (Manchester: Manchester University Press).

Cornwall County Council, 1952. *Town and Country Planning Act 1947. Report of the Survey (Part Two)*, (Truro: Cornwall County Council).

Cornwall County Council, 1976. *County Structure Plan. Topic Report: Employment, Income and Industry.* (Truro: Cornwall County Council).

Crine, S. and Playford, C., 1981. *Low Pay and Unemployment in Cornwall* (London: The Low Pay Unit).

Department of Employment, 1983a. *Employment Gazette,* (various issues).

Department of Employment, 1983b. *New Earnings Survey 1982*

Ennew, J., 1980. *The Western Isles Today,* (Cambridge: Cambridge University Press).

Fothergill, S. and Gudgin, G., 1982. *Unequal Growth,* (London: Heinemann).

Gilligan, J.H. (Forthcoming). Doing a community study, in Brown, P. and Gilligan, J.H. (eds.) *Doing Postgraduate Social Research,* (Swansea: Waldon Press).

Hechter, M., 1975. *Internal Colonialism,* (London: Routledge & Kegan Paul).

Henry, S., 1982. The working unemployed: perspectives on the informal economy and unemployment, *Sociological Review,* 30 (NS), pp. 460-477.

Massey, D., 1979. In what sense a regional problem?, *Regional Studies,* 13, pp. 233-243.

Massey, D., 1983. Industrial restructuring as class restructuring: production decentralisation and local uniqueness, *Regional Studies,* 17, pp. 73-89.

McNabb, R, and Woodward, N., 1979. *Unemployment in West Cornwall,* (London: Department of Employment, Research Paper no. 8).

Ministry of Agriculture and Fisheries, (annual). *See Fisheries Statistical Tables,* (London: HMSO).

Murray, J., 1865. *Murray's. A Handbook for Travellers in Devon and Cornwall,* (London: John Murray). 6th edition, revised.

Newby, H., 1982. *The State of Research into Social Stratification in Britain,* (London: Social Science Research Council).

Page, J.L.W., 1897. *The North Coast of Cornwall,* (Bristol: Crofton Hemmens).

Pearse, R., 1963. *The Ports and Harbours of Cornwall,* (St Austell: H.E. Warne).

Perry, R.W., 1983. *The Impact of Counterurbanisation on Cornwall,* South West Papers in Geography, no. 4.

Rallings, C.S. and Lee, A.N., 1978. *Cornwall: the 'Celtic fringe' in English politics,* (Paper presented to the BSA Sociology of Wales Conference, Gregynog).

Shaw, G. and Williams, A.M. (eds.) 1981. *Industrial Change in Cornwall,* South West Papers in Geography, no. 1.

Shaw, G. and Williams, A.M., 1982. *Economic Development and Policy in Cornwall,* South West Papers in Geography, no. 2.

Simmons, J., 1959. South Western v. Great Western: railway competition in Devon and Cornwall, *Journal of Transport History,* 4, pp. 13-36.

Symes, D.G., 1981. Rural Community Studies in Great Britain, in: Durand-Drouhin, J.L. and Szwenbrug, L-M. (eds.) *Rural Community Studies in Europe, Volume One,* (London: Pergamon).

Thompson, P. 1983. *Living the Fishing,* (London: Routledge & Kegan Paul).

Tunstall, J., 1962. *The Fishermen,* (London: MacGibbon & Kee).

Urry, J., 1981. Localities, regions and social class, *International Journal of Urban and Regional Research,* 5, pp. 455-474.

Williams, R.A., 1973. *The London and South Western Railway. Volume Two,* (Newton Abbot: David & Charles).

SECTION II

AGRICULTURE, LAND AND CAPITALIST PRODUCTION

6. Agrarian class structure and family farming

MICHAEL WINTER

INTRODUCTION: FAMILY FARMERS AS A CLASS

In recent years the study of agrarian class structures has
received a considerable fillip from a number of developments
in what has come to be known as the sociology of agriculture.
The new approach is far removed from the community studies
of agricultural areas which dominated much of the rural
sociological research agenda in Britain in the 1950s and
1960s. Carter (1975) reminds us of the inadequacies of
many of these earlier studies concerned, not so much with
agriculture but with the study of kinship relations from a
strongly functionalist, anthropological perspective. Such
a perspective could confine 'class' to manifest social
friction or consciousness *within* a locality, ignoring the
wider manifestations of class relations in the society
as a whole. Thus Williams (1963) commented on the absence
of 'class feeling' in his study area which was dominated
by family farming.

What was lacking was a clear political economy of
agriculture, which Carter (1979) takes to mean the social
relations of production. Similarly Newby (1977; 1978b)
locates his interests in rural society within an understanding
of social stratification and property relations in society
at large. Both writers, through their empirical research
as well as in their theoretical concerns, stress the need
to avoid studying agriculture and rural communities in
isolation. A more holistic approach is called for which
locates agriculture, and social relations within agriculture,
in the wider capitalist society.

So to turn to the study of family farming again is not
to seek a traditional way of life, an unchanging rural
milieu, but rather to study a particular form of capitalist
development in agriculture, indeed one of the peculiarities
of capitalist development in agriculture which Newby (1978a)
has suggested provides "at least a basis on which rural
sociology might proceed in the future". It is also to
engage in much wider debates about the nature of peasant
economy and society, the nature of pre-capitalist modes of

production, and the sources of uneven development. This
paper critically reviews strands of thought within these
wider concerns of sociology which offer insights into the
reasons for the survival of family farming within advanced
capitalist society.

Crucial to such an endeavour is an understanding of
the notion of family farming and its use as a category in
the analysis of agrarian class structures. This necessitates
a descriptive and empirical approach to family farming, as
well as theoretical elaboration of the concept. Descript-
ively, family farming refers to farming where the majority
of the labour as a direct input to farm production is
provided by the farmer and other members of the family.
The typical family farmer then occupies an atypical position
within capitalist society in that he is both owner of the
means of production and a direct producer. The functions
of management, ownership of capital, ownership or effective
control (in the case of tenants) of land, and provision of
labour are provided by the farmer and his family. Such
farms are likely to be on relatively small acreages,
employing relatively small quantities of capital and
labour.

The fundamental class cleavage between labour and
capital is not present *within* a family farming enterprise.
Nevertheless this is not to imply that social relations of
production are unimportant in family farming. Indeed it is
such relations internally within the farm family, and
externally between the family farm and other groups in the
economy and society - that are taken to be of crucial
importance in Friedmann's pioneering work on the theoretical
status of family farming (1978; 1980; 1981). Her attempts
to elaborate appropriate concepts for the analysis of
agrarian formations, although clearly within a Marxist
tradition, make considerable advances on previous Marxist
formulations which relegate family producers to a transitional or marginal class position. Poulantzas's characterisation (1978) is typical of the latter tradition in
describing such producers as a traditional petty bourgeoisie
which "does not belong to the capitalist mode of production,
but to the simple commodity form which was historically
the form of transition from the feudal to the capitalist
mode".

Other writers, conscious perhaps of the resilience of
family farmers, have characterised them as a unique and
unitary class, based on a distinctive, peasant or independent
mode of production. Though such an approach, particularly
as employed by Chayanov (1968), may be rich in insights
into the internal workings and economy of the family farm,
it fails ultimately to tackle the question of wider class
relations and overlooks the fact that family farming in
its modern form is quite specific to capitalism. Ennew,
Hirst and Tribe (1977) discussing the Marxist lineage of
this approach conclude that 'in no way can a 'theory of
peasant production' or even the elements of a 'peasant mode

of production' be extracted from the writings of Lenin and
Kautsky on the agrarian economy". The logic of this is
that we must look beyong kinship and family characteristics
in order to find the key explanatory features of a particular mode or form of production:

> Kinship does not determine the *formation*
> of 'familial' units of production, it
> is economic conditions independent of
> these relations which determined whether
> the families so formed are units of
> production in addition to units of
> consumption and whether the units of
> production have possession of the
> means of production. (My emphasis).
> (Ennew *et al.* 1977)

Analysis of family farming must consider the role of that
enterprise in the wider economy, and historical changes
particularly in the transition from feudalism to capitalism.
By emphasising links with the wider society through
commodity exchange we separate the family farm analytically
from the subsistence peasant producer. This is not to
imply that production under feudalism is inevitably for
direct consumption. Indeed, the production of commodities
was strongly developed under feudalism but was subordinated
to "the goal of feudal consumption" (Banaji 1976a) based
on "politico-legal relations of compulsion" (Anderson 1978).
Capitalist commodity production, and therefore family farm
production within capitalism, is for profit, accumulation
and/or reproduction via exchange rather than for mere
consumption.

It is the internal and external characteristics of
family farm production that are crucial. Friedmann calls
for the surrender of the term 'peasant' to give way to the
analytic specification of forms of production based on
internal characteristics of the unit and the external
characteristics of the social formation (1980) or, as used
in a later paper (1981), the mode of production. Thus
she posits a 'double specification', whereby the two sets
of characteristics "determine the conditions of reproduction
of the form and the manner in which its circuits of reproduction intersect with those of other classes" (1980).
There is nothing in Friedmann's work to imply that the
precise empirical *conditions* of reproduction are necessarily
given in the analytic notion of the doubly specified simple
commodity form of production. Her approach allows for
heterogeneous conditions of reproduction within a single
economy based on contrasting internal relations, differential
relations with capital or the state, and different tenurial
or product-market arrangements.

Crucial to Friedmann's approach is the notion of
commoditisation as a process within capitalism. This is
"a logical concept, referring to the complete separation
of the household from all ties except those of the market"
(1980). The end point of commoditisation, in terms of

family farming, is where all relations (except that of
labour) are fully commoditised - the pure notion of simple
commodity production. In reality such a pure form rarely
exists and most family farmers offer various forms of
resistance to the full logic of commoditisation, and in a
number of ways resistance to commoditisation can be seen
as a strategy for survival under certain conditions. Much
of the rest of this paper will critically examine attempts
to explain the survival of family farming, in the light
of the theoretical position briefly sketched above.

THE SURVIVAL OF FAMILY FARMING

It is the peculiar nature of land as a factor of production
in agriculture that is at the root of the survival of
simple commodity production. Agriculture's dependency on
land has both inhibited capital penetration and allowed
a peculiar development of a highly technological industry
sub-divided into relatively small units.

     As Ricardo and Marx pointed out, land is relatively
fixed in quantity and cannot be increased in the short
term or transferred from one location to another. Lenin
(1956) also commented on this monopoly of land and its
effects on the capital penetration of agriculture:

                agriculture possesses a special monopoly
                which is peculiar to it, which is unknown
                in industry, and which cannot be eliminated
                under capitalism, viz. the monopoly of
                land. Even if there is no private
                ownership of land ... the very possession
                of land, its occupation by individual,
                private farmers creates a monopoly.
                In the principal regions of the country
                all the land is occupied, and an
                increase in the number of agricultural
                enterprises is possible only if the
                existing enterprises are parcelled
                out into smaller ones; the unimpeded
                creation of new enterprises side by side
                with the old ones is impossible. The
                monopoly of land is a brake, which
                retards the development of agriculture,
                retards the development of capitalism
                in agriculture.

     Not only does this natural monopoly of land hinder the
creation of new agricultural enterprises, it also limits
the expansion of existing holdings. Whereas in most forms
of manufacturing industry, capital accumulation can occur
independently of the centralization and concentration of
land, in agriculture large holdings can only be formed
through the amalgamation of several smaller holdings.
Chayanov rather graphically summed up the problem when he

pointed out that men cannot gather the sunbeams that fall
on 100 hectares and apply them to just one hectare (Harrison
1977). Farmers who wish to accumulate and concentrate are
dependent on the land market. It is one of the contradictions of capitalism that private property, integral to
capitalist expansion and accumulation should in the case of
agriculture be a barrier to capitalist penetration. Land
constitutes a property monopoly (Goss *et al.* 1980), as
commented on by Kautsky:

> bourgeois property recognizes only one
> basis for expropriation - default. As
> long as the peasant repays the capitalist
> and the state, his property is sacrosanct.
> This poses a serious obstacle to the
> growth of big landed properties. The
> process is most difficult where small
> property predominates exclusively.
> (Banaji 1976b)

Where land ceases to be such a determining factor of agricultural production due to technological advances, as in the
pig and poultry sectors of recent years, there may be
dramatic capital penetration and re-structuring in the
industry.

It is not only that agriculture's dependence on land
slows down the rate of centralization and concentration,
such a use of space also creates major differences between
large-scale production in industry and agriculture, a
point again developed by Kautsky:

> The expansion of an industrial enterprise
> signifies a growing concentration of
> productive forces with all the concomitant
> economies - of time, costs, material
> management, and so on. In agriculture,
> on the other hand, the expansion of a
> given enterprise on the same technical
> basis amounts to a mere extension of the
> area under cultivation, and thus entails
> a greater loss of material, a greater
> deployment of effort, resources, time,
> for the transport of material and men.
> (Banaji 1976b)

Such diseconomies of distance, which vary according to
levels of technology, soil fertility and enterprise type,
are linked to economies of scale also difficult to achieve
in agriculture. Certainly, for the pastoral farming which
is the predominant mode of family farming in Britain, the
limits of economies of scale are achieved with relatively
low levels of capital concentration. A study of farm size
and efficiency by Britton and Hill (1975) has found that
"the acreage beyond which significant improvements in
efficiency did not appear to occur was 100-150 acres for
dairy farms, 150 acres for mixed farms, (and) 200-250 acres
for cropping farms".

Furthermore, where diseconomies are suffered by smaller farms, it is more likely to be due to labour surplus and resulting underemployment on small farms, than to inadequate technology - problems increasingly overcome through farmers' involvement in outside, part-time employment or involvement in farm tourism enterprises. Of course this particular aspect of agriculture's use of space does not remove the incentive for the accumulation and concentration of capital within agriculture, rather it makes the process more difficult, for family farmers can frequently compete on equal terms with capitalist producers.

As a result of this competition, family farming has, in some ways strengthened its position in twentieth century Britain. Concurrent with a marked increase in the owner-occupation of farms has been a steady drop in the average size of the farm labour force on British farms. At the turn of the century probably less than 15 per cent of agricultural labour input in Britain was provided by the farmer and his family, the rest being provided by mixed labour. Since that time, and especially since 1945, there has been a slow increase in the size of farms and scale of production but this has not been sufficient to prevent the massive decline in numbers of agricultural workers consequent on the introduction of labour-saving technology. Nowadays, something in the order of 60 per cent of farm labour nationally is provided by the farmer and his family, and in the pastoral west the figure can be as high as 80 per cent in some counties. The nature of land as a factor of production is undoubtedly a major reason for the survival, indeed resurgence of family farming.

AGRICULTURAL RENT

Another crucial component of the peculiarities of capitalist development in agriculture, based on the natural and private monopoly of land, is the *rent relation* in agriculture. Theories of rent, in particular interpretations of Marx's theory, have been keenly and closely argued in recent years, and the subject is highly complex. What is important here is to pick out those aspects of rent theory that have a direct bearing on the survival of family farming. It is important to treat rent not as an arbitary financial category - a 'production cost' in the terminology of liberal economies - but as a relational and distributional expression of the production of surplus value in agriculture, dependent as it is on the private monopoly of land. As Fine (1979) puts it, "Marx's rent theory is concerned with the question of how the laws that apply to industrial capital in general are modified by the existence of landed property".

Capitalist ground-rent (it must be categorized as such to distinguish it from rent under conditions of feudalism) has two components. The first, *differential rent,* has to be seen in the context of competition within the agricultural

sector (Fine 1979). Differential rates of profit, periodic
surplus profits, exist in all industries, but outside
agriculture competition mechanisms and the flow of capital
serve to equalise rates of profit over time. In agriculture
this does not occur - hence the existence of differential
rent:

> Differential rent ... is nothing but the
> excess profit yielded by capitals employed
> in above-average conditions owing to the
> (establishment of) one identical market-
> value in every sphere of production. This
> excess profit consolidates itself only in
> agriculture because of its *natural basis*
> and, furthermore, the excess profit flows
> not into the pockets of the capitalist but
> into that of the landowner since it is the
> landowner who represents this natural basis.
> (Marx 1975, vol. II)

Differential rent is usually based on natural variations
in productivity arising from soil fertility and climate,
although variations due to differential capitals are also
important.

The other component of capitalist ground-rent - *absolute
rent* - is based, not on differential productivities of
capital, but on the barrier posed by landed property whereby
"the market-price must rise to a level at which the land
can yield a surplus over the price of production, i.e. yield
a rent." (Marx 1974, vol. III) Fine explains the difference
between differential rent (DR) and absolute rent (AR) in
the following terms:

> Both concern the obstacle to capital
> investment posed by landed property and
> the associated appropriation of surplus
> profit in the form of rent, but each is
> located at a different level of analysis
> and therefore has a different source of
> surplus. DR depends upon the divergence
> between individual and market values,
> AR on the divergence between market values
> and prices of production.
> (Fine 1979, p. 258)

Before we consider rent as a barrier to capitalist
penetration, we need to establish the heuristic status of
rent, even under owner-occupation. This is important in
the context of the present debate as owner-occupation has
come to dominate West European agriculture. In Marx's
words "the price of land is nothing but capitalised and
therefore anticipated rent" (1974, vol. III). So as with
the payment of ground rent, accumulation is affected:

> interest on the price of land - which
> generally has to be paid to still

another individual, the mortgage
creditor - is a barrier.
(Marx 1974, vol. III)

Marx does not dwell at length on the *long-term* consequences
of owner-occupation. However Murray has emphasised the
difficulties caused by capitalised rent to incoming farmers,
and the partial insulation from true market laws of those
owner-occupiers who have inherited holdings, whose capitalised
rent was paid in a sale price many years previously:

When owner occupation is established
through the market, the farmer is from
the first burdened with debt and new
investment is restricted ... On the
other hand an owner occupier free from rent
is like-wise insulated (to the extent of
his differential rent) from the law of
value. We find the land parcelization and
bad husbandry that characterizes many
peasant societies. For this reason capital
has often ensured that new owner occupiers
are from the first dependent on credit
for the payment of land purchase price.
(Murray 1978)

Thus capitalised ground rent has a double effect on
capital accumulation in agriculture. On the one hand,
those farmers who have long been owner-occupiers can retain
the capitalised rent, or surplus profits, and are thus
partially insulated from the law of value. Murray's
comments on peasants could equally apply to some of the
small family farmers of the pastoral areas of West Britain.
Such farmers who are not having to 'pay' for their land,
either through rent or mortgage payments, frequently farm
at lower levels of capitalization than they would otherwise
have to. By avoiding the full rigours of market competition
such farmers are buttressed against capital penetration
into agriculture. Secondly where ground rent is extracted,
through mortgage payments, there is similarly a barrier to
capital penetration - the premature equalization of the
rate of profit.

A *general* theory of rent of this nature points to a
number of barriers to capitalist penetration of agriculture.
It does, of course, cover a number of different *social*
relations which may emerge under different tenurial arrangements. It is also an example of how family farming can
flourish in an unevenly developed, commoditised market.
Simple commodity production, as outlined by Friedmann (1981),
presupposes full mobility of land, and full ties to the
land market. Clearly Murray's 'peasants' are in a position
of resistance to this fully developed feature of commoditisation, unlike the mortgages and tenants 'paying' for
their land. Both positions can provide the conditions for
the survival of family farming, but both imply contrasting
internal and external social and economic relations. This

is of fundamental importance for two reasons. First, it provides a research agenda for empirical questions concerning family farming in the agrarian class structure. For example what are the contrasting relations likely to be found within the family at different levels of land commoditisation? How are relations with landlords and suppliers of credit and inputs likely to be affected?

Secondly, contrasting degrees of resistance to commoditisation contribute to variation in the conditions for the survival of family farming and the social relations engendered. This should lead us to be suspicious of any single theory which offers a general explanation of family farm survival.

PRODUCTION TIME AND LABOUR TIME

There is another area of economic analysis based on agriculture's peculiar use of space which needs stressing in a discussion of the survival of family farming in advanced capitalist society. This concerns the difference between *production time* and *labour time,* and problems of the *circulation of capital* in agriculture. Marx stressed that production time consists of two parts: one period when labour is actually applied in production, and a second period when the unfinished commodity is "abandoned to the sway of natural processes" without being at that time in the labour process (Marx 1974, vol. II; Mann and Dickinson 1978). The implications of this for agricultural production are important. Because of agriculture's peculiar use of space, in particular the continuing dependence on the cycle of the seasons, its production time remains considerably greater than its labour time. Linked to this is the fact that capital can only circulate slowly where there is a lengthy production time. The speedier the movement of capital the higher total profits are likely to be. Thus capital is attracted to those industries where, unlike agriculture, the production time can be shortened. The possible effect of this on the development of capitalism in agriculture is important:

> It is our contention that the capitalization of agriculture progresses most rapidly in those spheres where production time can be successfully reduced. Conversely we maintain that those spheres of production characterized by a more or less rigid non-identity of production time and labour time are likely to prove unattractive to capital on a large scale and thus are left more or less in the hands of the petty producer.
> (Mann and Dickinson 1978)

Empirically, however, this argument is only partially sustainable. Whereas capitalisation was proceeded rapidly in the pig and poultry sector where production time has been shortened, the process is less pronounced in dairying where the production time is also relatively short. The argument has to be seen in the context of other factors, particularly the greater importance of land in some branches of agricultural production, such as dairying, than in others, such as pigs and poultry.

ARTICULATION

Another way in which space is of crucial importance when considering reasons for the survival of family farming concerns the spheres of *uneven development* and *articulation of modes of production,* in other words the ways in which capitalism itself finds *spatial expression.* When we talk about the peculiar nature of land as a factor of production (and the corresponding implications of land monopoly, rent, and production and labour time differences), we are considering the implications of *agriculture's* peculiar *use* of space. Now we must consider agriculture *in* space, in other words the effects on agriculture of *capitalism's* use of space. These are factors located not in agriculture itself but in other, wider relationships. The point is best made when considering differences in the agriculture of different regions of a country:

> Even in those few areas where particular
> branches of production *have* entirely
> dominated the economy, it is not possible
> simply to assume that such areas will be
> the same as others equally so dominated.
> It is more that family farming is the most
> successful form of production for putting
> the maximum volume of surplus peasant
> labour at the disposal of urban capitalism.
> It also constitutes the most efficient way
> of restraining the prices of agricultural
> products ... The 'longevity' of family
> farming in contemporary capitalism is not
> difficult to explain. It follows from the
> facility and the rapidity with which the
> family productive unit adapts itself to
> the requirements of the urban system.
> (Vergopoulos 1978)

The small farmer is, in effect, reduced to the state of proletarian by the extraction of surplus value by urban capital through the domination of agriculture by large oligopolistic or monopoloistic agri-business firms.

The functionalist perspective is expanded by Banaji (1977):

> the relations of production which
> tie the enterprise of small commodity

> producers to capital are already
> relations of capitalist production.
> Between the market and the small producer,
> capital intervenes with the determinate
> forms and specific functions of both
> merchant and industrial capital ...
> The social process of production incorporating
> the immediate labour-process of the small
> peasant enterprise is governed by the
> aims of capitalist production.

Functionalist articulation can also occur at the political level; Bernier (1976) has noted how small farmers can be used as a political tool of the bourgeoisie against the working class. But there are various problems associated with these ideas when applied to advanced capitalist societies, as already indicated by our discussion of Friedmann. There is the tendency, particularly noticeable in Vergopoulos, to resurrect dualism and a false analytical separation of country and city. In Britain at least there are strong reasons for speaking of an early developed *agrarian* capitalism from which industrialism grew (Merrington 1978). Another danger in the approach is to imply that, because a functional relationship prevails, family farms are necessarily reproduced by capital, a crude descent into dependency theory. Ultimately, family farmers and peasants reproduce themselves through their own labour. Articulation of modes or forms of production to be of any theoretical use has to refer to a relational rather than a functionalist concept. "The question is how the conditions of production and reproduction are determined by the operations of capital and of the state" (Bernstein 1977). A theory of functionalism which is too deterministic fails to appreciate internal differentiation and change within the agrarian sector.

Nevertheless, it is undoubtedly true that family farming tends to dominate agriculture on the periphery of the capitalist state, areas that have been traditionally associated with lack of development and incomplete penetration by outside capital. Certainly in areas of low wage levels and high unemployment there is little incentive for small farmers of farmers' sons to give up farming. Distance from the centres of economic activity implies also distance from political and ideological processes of the central state, encouraging forms of local consciousness and ideology that buttress and are supported by petty bourgeois ideology.

CONCLUSION

This paper has necessarily been incomplete in some sections, in that it has tried to cover a wide range of ideas concerning the survival of family farming in order to illustrate both the complexity of the subject and the inadequacies of taking any one particular argument on its own. Nevertheless, there have been omissions, due in part to the chosen focus of the paper - the family farm in *space*. Underlying *economic* determinants have been discussed but considerations of the ideological and political levels have been neglected, not least the role of the State in the survival of family farming, and of course the ideologies of farmers themselves. In the final analysis, family farmers can only survive through self-reproduction, through their own labour which they must rationalize ideologically, and justify politically.

To draw together the strands of the different ideas concerning the survival of the family farm is difficult and only a few beginnings can be hinted at here. It is apparent that the various ideas presented in the paper are in themselves incomplete explanations for the survival of family farming. At the most simple level, it is plain that the rent theories, concentrating on the peculiar nature of land in agricultural production, are positing barriers to an otherwise *willing* capitalism. Landed property is a barrier to capital penetration and accumulation in agriculture: capital would penetrate agriculture if only it could! Conversely, the argument about the non-identity of production time and labour time and corresponding slow rates of capital circulation in agriculture hypothesizes that capital would not wish to penetrate agriculture; agriculture is not sufficiently attractive to it.

The arguments, both based on different facets of agriculture's use of space appear to be mutually exclusive: offering opposite conclusions from considerations of the same constraining factor - land. But the arguments are based on too rigid a delineation of the spatial character- istics of agriculture, ignoring changes over time and differences in farming patterns and enterprises, i.e. the complex, historically and spatially specific, articulation that the family form of production has with capitalism. Thus one of the main differences in the two arguments concerns, *not* conflicting interpretations of the constraints of land in agriculture, but the emphasis placed on different agents in agriculture. The production time/labour time argument tends to hypothesize a static agriculture unattrac- tive to *outside* capital, whereas the land monopoly argument tends to hypothesize an agriculture in which the *internal* agents of change are stifled. In other words, the difference is not over the nature of land in production, but over an understanding of how agricultural developments take place and the relations between town and country. The production time/labour time argument is in danger of reviving a dualism inappropriate to most advanced capitalist formations, by positing an *urban* capital seeking to transcend pre-capitalist or non-capitalist structures through the total subsumption of labour to capital. The argument becomes similar to some of the functionalist articulation arguments, through the false analytical separation of town and country, of urban industrial capital and a rural peasantry. But capital is not an amorphous body acting from outside agriculture: some of the basis for agrarian change and agrarian capitalism can be found *within* agriculture. Thus the land monopoly argument rests on the difficulties that are faced by *capitalist or petty capitalist farmers,* as well as outside capitalist seeking to penetrate agriculture. The incentive for capitalist development is from without *and* within agriculture; indeed, some of the very seeds of the dissolution of the family form of production can be seen in the family farmer's own desires for expansion.

The different arguments are not necessarily contradictory but are based on analyses of different agricultures at different times. What is needed is concrete research to establish which agents of capital, fractions of capital, are active within any particular process of production in agriculture at any one point in time. We need to analyse changing relations between agricultural capital and finance and industrial capital to analyse the role of farmers' political groups in state political processes. The need is for comparative studies of agriculture to see how a particular form of production varies over time and space. Friedmann's concepts of commoditisation and a double specification of the family farm of production offer a particularly helpful way forward for this analysis.

REFERENCES

Anderson, P., 1978. *Passages from Antiquity to Feudalism,* (London: Verso).

Banaji, J., 1976a. The peasantry in the feudal mode of production: towards an economic model, *Journal of Peasant Studies,* 3, pp. 299-320.

Banaji, J., 1976b. Summary of selected parts of Kautsky's *The Agrarian Question, Economy and Society,* 5, pp. 2-49.

Banaji, J., 1977. Modes of production in a materialist conception of history, *Capital and Class,* 3, pp. 1-44.

Bernier, B., 1976. The penetration of capitalism in Quebec agriculture, *Canadian Review of Sociology and Anthropology,* 13, pp. 422-434.

Bernstein, H., 1977. Notes on capital and peasantry, *Review of African Political Economy,* 10, pp. 60-73.

Britton, D.K. and Hill, N., 1975. *Size and Efficiency in Farming,* (Farnborough: Saxon House).

Carter, I., 1975. The peasantry of northeast Scotland, *Journal of Peasant Studies,* 3, pp. 145-191.

Carter, I., 1979. *Farm Life in Northeast Scotland,* (Edinburgh: John Donald).

Chayanov, A.V., 1968. *The Theory of Peasant Economy,* (Illinois: Richard D. Irwin).

Ennew, J., Hirst, P. and Tribe, K., 1977. 'Peasantry' as an economic category, *Journal of Peasant Studies,* 4, pp. 295-322.

Fine, B., 1979. On Marx's theory of agricultural rent, *Economy and Society,* 8, pp. 241-278.

Friedmann, H., 1978. World market, state and family farm: social basis of household production in the era of wage labour, *Comparative Studies in Society and History,* 20, pp. 545-586.

Friedmann, H., 1980. Household production and the national economy concepts for the analysis of agrarian formations, *Journal of Peasant Studies,* 7, pp. 158-184.

Friedmann, H., 1981. *The family farm in advanced capitalism: outline of a theory of simple commodity production in advanced capitalism,* (paper prepared for the thematic panel 'Rethinking domestic agriculture', American Sociological Association, Toronto).

Goss, K.F., Rodefeld, R.D. and Buttel, F.H., 1980. The political economy of class structure in US agriculture: a theoretical outline, in: F.H. Buttel and H. Newby (eds.) *The Rural Sociology of the Advanced Societies: Critical Perspectives,* (London: Croom Helm).

Harrison, M., 1977. The peasant mode of production in the work of A.V.Chayanov, *Journal of Peasant Studies,* 4, pp. 323-336.

Lenin, V., 1956. *Capitalism and Agriculture,* (New York: International Publishers).

Mann, S.A. and Dickinson, J.A., 1978. Obstacles to the development of a capitalist agriculture, *Journal of Peasant Studies,* 5, pp. 466-481.

Marx, K., 1974. *Capital,* (London: Lawrence and Wishart).

Marx, K., 1975. *Theories of Surplus Value,* (London: Lawrence and Wishart).

Merrington, J., 1978. Town and country in the transition to capitalism, in R.H. Hilton *The Transition from Feudalism to Capitalism,* (London: Verso).

Murray, R., 1978. Modern landed property and rent, part 2, *Capital and Class,* 34, pp. 11-33.

Newby, H., 1977. *The Deferential Worker: a Study of Farm Workers in East Anglia,* (London: Allen Lane).

Newby, H., 1978a. The rural sociology of advanced capitalist societies, in H. Newby (ed.) *International Perspectives in Rural Sociology,* (Chichester: John Wiley & Sons).

Newby, H., Bell, C., Rose, D. and Saunders, P., 1978b. *Property, Paternalism and Power: class and control in rural England,* (London: Hutchinson).

Poulantzas, N., 1978. *Classes in Contemporary Capitalism,* (London: Verso).

Vergopoulos, K., 1978. Capitalism and peasant productivity, *Journal of Peasant Studies,* 5, pp. 446-465.

Williams, W.M., 1963. *A West Country Village Ashworthy: Family, Kinship and Land,* (London: Routledge & Kegan Paul).

7. Land ownership and farm organisation in capitalist agriculture

TERRY MARSDEN

   A major characteristic of agricultural development in Britain over the last hundred years has been the twin processes of accumulation and concentration of capital and wealth.   The study of British agriculture in this sense becomes a study of capital accumulation.

   Changes in the size of farms are a measure of such concentration.   There was a loss of some 275 000 holdings in Britain between 1908 and 1975, a reduction of 54 per cent.   As a consequence the average area of holdings rose from 25 ha in 1909 to 50 ha in 1977.   During the more recent period, 1965-75, while there was a net loss of some 22 000 small holdings (i.e. farms providing work for one or two full-time men), the proportion of the total agricultural area occupied by holdings of 120 ha and above increased from 33 per cent to 43 per cent.   Even where isolated pockets of small-scale farming exist, the domination of markets by the larger producers as well as the increasing capital investments required to sustain viability, ensure that the small producers occupy a diminishing role in terms of national output and number of holdings occupied.

   This paper attempts to demonstrate how the structure of landownership and farm businesses have adapted to the needs of this capitalist process by reference to evidence from eastern lowland Britain - one of the most advanced agricultural regions in Europe.   This will entail, first an evaluation of the development of landownership and farm organisation during the periods of high farming in the nineteenth century, and then a discussion of the characteristics of modern farm organisation.

THE PROCESS OF ACCUMULATION AND CONCENTRATION

The macro socio-economic circumstances which now bear upon farmers and landowners have often been likened to a treadmill (Cochrane 1958; Dexter 1977), whereby increasing capital inputs and business size are made imperative in

order to reduce costs of production and maintain incomes in
an environment where market demand is at best stable or
dwindling. The consequent problems of over-production,
aided and abetted by government and EEC grant aid, only
serve to exacerbate the problem, forcing farmers into
another vicious spiral of capitalisation. Some of the
economic and geographical implications of these processes
have been examined by Britton and Hill (1975). They have
identified the increasing thresholds in terms of the size
of operations needed to ensure viability. Because of the
economic limitations imposed on small farms and the increased
land-using capacity (through mechanisation) of each farm
employee, there is great pressure to establish larger units.
Where 'efficiency' is measured by output per £100 of inputs,
including family labour, Britton and Hill estimated that
farms of less then 600-800 standard man days (or farms
approximating to 110-150 acres) are shown to be less
efficient than farms above that size. Such economies of
scale seem to operate up to farms of 300 acres (120 ha),
beyond which the advantages of increasing scale seem to
reach a plateau. The exact point at which the economies
of scale evaporate is difficult to estimate either in terms
of size or enterprise combinations. What is clearer is
that very large farms have increased in number. Over the
period 1968-75 the number of holdings over 1000 acres rose
steadily from 10.1 to 14.1 per cent.

    While this has meant a considerable reduction and amal-
gamation of holdings in the traditional small-scale farming
areas of western, upland Britain, in the already more
prosperous agricultural counties of eastern England such
forces have promoted the further development of large-scale
farming, leading to increased incomes and the further
concentration and accumulation of capital. Hence, for the
large-scale, arable dominated areas, the treadmill has been
extremely beneficial, facilitating increased production and
the further exploitation of scale economies.

    The cycle of accumulation has therefore been one
encompassing the processes of production and the marketing
or exchange of products (Figure 7.1). The essential
elements in the simplified model outlined here are that
changes in the scale and types of production have influenced
the structure of landownership and the business and labour
organisation of farms. They have satisfied the demands of
the market and been encouraged by state intervention.
Increasing incomes and profits have enabled changes in the
methods of production and organisation. As it will be
argued, this has produced a more complex landowning structure,
with many farm businesses having acquired extra parcels
of land to develop new enterprises.

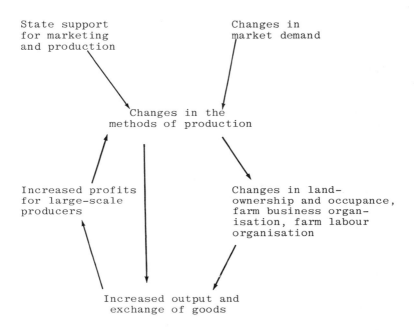

Figure 7.1   The cycle of accumulation

LANDOWNERSHIP AND OCCUPANCY IN THE NINETEENTH CENTURY

In North Humberside the traditional three-tier hierarchy
of landowner, farmer and worker prevailed until the end
of the nineteenth century. Large landed estates had been
established by a small number of families. Table 7.1 shows
that 19 such families with estates above 2500 ha controlled
nearly 37 per cent of the land. Their social origins
varied from ancient landed families; to the more recently
ennobled merchant families of the seventeenth and eighteenth
centuries who had converted their wealth into property and
social standing; and to the *nouveau riche* who had gained
substantial commercial fortunes in the earlier part of the
nineteenth century. There was a larger number of small
landowners - the squirearchy - with estates of under
1250 ha.

     This is in keeping with the claim of Whetham (1978)
that, for Britain as a whole, fewer than two thousand
families controlled more than 14 million acres of land in
1873. Caird, writing at the zenith of the landed estates,
proclaimed:

Table 7.1   Major landowners in East Yorkshire, 1873

	Over 10 000	7500- 9999	5000- 7499	2500- 4999	1250- 2499	500- 1249
Number	2	3	3	11	14	41
Hectares	25 521	24 315	16 815	38 912	23 535	32 080
% total area	8.92	8.50	5.89	13.52	8.22	11.21
Cumulative %	8.92	17.41	23.29	36.81	45.04	56.25

> In short our system is that of large
> capitalists owning the land, of smaller
> capitalists each cultivating five times
> more of it than they would have means to
> do if they owned their farms, and of
> labourers free to carry their labour to
> any market which they consider most
> remunerative.
> (Caird 1851, p. 67)

In North Humberside an apparently prosperous tenantry
had developed on large farms (average area 102.8 ha in 1851)
particularly on the light chalk land of the Wolds.   To
judge from contemporary accounts, however, the relationships
between landlord and tenant were tenuous and insecure, with
many tenants being without a written agreement.   A common
custom was yearly tenancies terminable at six months notice.
Although it is difficult to measure the turnover of tenancies
during this period, circumstantial evidence from the 1851
census enumeration books indicate considerable mobility
among the tenantry.   For instance, in a sample of farms
covering both the Wolds and Holderness regions, three-
quarters of all heads of households were born outside the
township of their existing tenancy, and the birthplaces of
their children often indicate one or more other places of
residence as well.

There is little evidence to suggest that they represented a stable class based upon generational continuity
and attachment to their farms.   Indeed if we examine the
family structures as revealed in the 1851 census, it seems
that a significant proportion of farm households were
either single persons (13.4 per cent), groups of unmarried
siblings (11.6 per cent) or apparently childless marriages
(8.7 per cent).   More than one-third of the holdings were
without a direct male heir present on the farm.   As a
consequence, it was rarely possible for the family to
generate a sufficiently large and constant supply of labour

to satisfy either the agricultural or domestic spheres of activity (see Table 7.2). Through the insecure, tenurial relations and their own biological inertia, the North Humberside tenantry was very much a weak link in the system which Caird elaborated.

Table 7.2   Number of Family Members working on the farm, 1851

Number of family members working on the farm	0	1	2	3	4	5	Average number of family members working on farm
Wolds	5	39	20	10	1	1	1.55
Holderness	5	59	14	5	0	1	1.27
Sunk Island	0	8	4	0	0	0	1.33
All farms	10	106	38	15	1	2	1.44

Source: Census Enumerators Books, 1851.

For the tenantry to make a profit above the rents paid to the landlords large proportions of hired labour were necessarily employed. Except for the very small farms hired labour was a necessity. Just as there was considerable diversity in landlord-tenant agreements, so too were there complex methods of employing and using labour on the farms. Three major sources of hired labour can be identified: *farm servants* who were hired for the year at Martinmas (November) and lived on the farm; *day labourers* from the village or occasionally from neighbouring villages; *occasional or seasonal* labour supplied by groups of women and children from the local area or sometimes gangs of migrant, Irish labour.

The number of farm servants tended to increase according to the size of farm, often rising to over a dozen on the large Wold farms. They came, for the most part, from the families of agricultural labourers in the surrounding villages and formed a structured workforce with the opportunity for promotion and increased wages based upon age and experience. The youngest were employed as ploughboys or general servants whose nondescript role might include work both indoors and out. Next in the hierarchy was the waggoner, responsible for teams of horses; and in charge of the entire complement was the foreman - normally the eldest on the farm. The shepherd occupied a special position on the Wold farm where the sheep were often the most important single enterprise.

Though the farm servants represented the stable workforce and provided experience and skill throughout the

the agricultural year, they were usually outnumbered by
the day labourers who were drawn from the local villages
(Table 7.3).

Table 7.3   Farm servants and day labourers working on
farms in East Yorkshire, 1851

Number of hired workers on farms	0	1	2-3	4-5	6-9	10+	Total	Average
Farm servants	24	24	50	31	31	14	646	3.76
Day labourers	20	30	48	28	22	23	719	4.18

Source: Census Enumerators Books, 1851.

The predominantly 'closed' villages of the Wolds limited
the accumulation of labour (and hence the poor rate liab-
ility). This meant that while a strong demand for day
labourers existed they had to be recruited over a wide area,
often necessitating a journey of several miles to and from
work each day. As an incentive farmers would sometimes
offer temporary sleeping places on the farm and provide
three hearty meals a day. Day labourers introduced an
element of flexibility into the labour supply; they were
hired on a daily or weekly basis or for whatever period
suited the particular seasonal task; and their hours of
work varied according to the season of the year. Many had
an opportunity to supplement irregular wages by cultivating
a small allotment in the village or grazing a cow in a
rented field or along the roadside verges.

     The third tier of workers - occasional and seasonal
labour - were necessary largely during the cereal harvest
when much greater labour inputs were needed. At such times
the entire resources of the farm household, including the
female servants who were not normally called upon to work
in the fields, were pressed into service together with the
village labourers.

     Several important points need emphasising about farm
organisation at this time. Firstly, there was a stable
landowning structure which had adjusted to the increasing
demands for agricultural produce by separating production
from ownership and establishing a complex system of tenurial
relations. Land not only enhanced economic wealth of course,
but for the burgeoning squirearchy in particular, it
provided a source of status. Rents were pitched at levels
which ensured that the tenantry sensitively geared their
production energies to the vagaries of market demand.
Strickland described the tenantry of the East Riding as:

          intelligent and liberal minded men who
          are ever ready to adopt any new system,

or to try any experiment in agriculture,
bearing with it a reasonable prospect of
success.
(Strickland 1812, p. 52).

At the same time the farm family, an essentially nuclear
family, was flawed by imperfections in its social and
biological mechanisms for it frequently failed to provide
a line of succession and it proved altogether inadequate
to the labour demands of large-scale, arable farming.
These demands were met through the employment of non-family
labour, and it is clear that the system of production
operating in East Yorkshire in the mid-nineteenth century
was dominated by the exploitation of hired labour.

FORCES OF CHANGE IN LAND AND CAPITAL

In the years since the mid-Victorian period of high farming
there have been profound changes to the structure of capitalist agriculture. The cyclic processes of production and
marketing outlined in Figure 7.1 have turned, generating a
different system of landownership and farm business organisation. Firstly, the processes of accumulation and
concentration of landownership which had characterised the
period from 1790 to 1870 were modified in the agricultural
depression which lasted, with a brief respite during the
First World War, from 1873 to 1939. Legislation giving
tenants security of tenure (particularly the Agricultural
Holdings Act of 1908), decreasing returns on fixed capital
and the gradual increase in the price of land with vacant
possession, all led to the steady disintegration of many
landed estates both in North Humberside and elsewhere after
1870, with the pace quickening in the years immediately
before and after the First World War (Sturmey 1955). Such
a diminution of control was complemented by an absolute
loss of power in political and economic life for the great
landowners. Many sitting tenants were encouraged to buy
their farms by taxation incentives introduced during the
latter part of the First World War. By 1919 new methods of
calculating land values were imposing heavier death duty
payments upon the landowners which further induced their
heirs to break up the estates. Though many aspiring owner
occupiers were to face bankruptcy in the inter-war years
due to the burden of mortgage payments, the trend towards
increasing owner occupation, and a reduction in the number
and power of landed estates has continued through to the
present day.

The depression of the 1930s brought a temporary halt
to these processes. The drop in demand for quality malting
barley and for heavy weight mutton - the two staple enterprises of the typical Wold farm - meant that the depression
sorely hit the East Riding farmer. As Ruston argued in
1936, it brought about

A more marked and more baffling state
of depression in the Wolds than elsewhere

135

in farming, since the system is based
almost entirely on the markets of
products which appear to be suffering
a more or less permanent eclipse.
(Ruston 1936, p. 83).

During this period and through to the early years of the
Second World War the increase in owner occupation and
dissolution of estates were halted. The security of tenure
conferred upon tenants in 1941 removed much of the incentive
for them to purchase their farms.

Despite the setbacks of the 1930s, by the early post-
war period the underlying processes of accumulation and
concentration re-emerged. There was a gradual increase in
the size of farms both in Eastern England and North
Humberside, a replacement of labour by capital as farmers
were finding it possible to increase production once more,
and a system of Government intervention after 1945 which
strongly favoured the larger more productive farms (Self
and Storing 1962, p. 34). Donaldson and Donaldson (1972)
estimated, for instance, that by the mid-1960s the arable
farm sector received three times more assistance (£180
million) than the livestock sector (£63 million). Market
trends and Government intervention encouraged farmers to
accumulate and re-invest capital, while the remaining
estate landlords experienced increasing taxation pressures.
By the mid-1970s the capitalist farm structure had changed
dramatically since the turn of the century. From being
based upon a landlord-tenant system many farmers were now
owner occupiers. Sturmey estimated in 1955 that 50 per
cent of farmers in Britain were owner-occupiers, and in
North Humberside, by 1973, 42 per cent of farms were owned
and another 25 per cent partly owned. As in most of lowland Britain such changes were associated with a specialisation and intensification of production. Aided by
technical advances which increased yields and introduced
new strains of cereals, North Humberside's barley acreage
reached a peak in 1967 and has since maintained its position
(in terms of acreage as a proportion of total tillage) as
Britain's pre-eminent barley-producing area. Where large
farms have developed without the former constraints of
rotations, intensive systems of livestock production and
the introduction of new cash crops (particularly vining
peas and oil seed rape) have superseded the 'barley-and-
sheep' character over much of the region.

LANDOWNERSHIP AND OCCUPANCY

The very strengths of the nineteenth century arable farming
system - the landlords and the agricultural workforce -
have both been attenuated. The farm family, however,
hitherto perhaps the weakest element in the system has developed a much more influential role. Today the farmer and
his family own more of the land they cultivate and provide
a much higher proportion of the labour resources. The

process of accumulation and concentration of farmland has
been renewed but with the farm family, and its often complex
farm business organisation as the central institution.
The study of agricultural change in lowland Britain is
therefore of a group in society who have considerably more
control of their own actions and the actions of others than
was the case a hundred, or even fifty years ago. Large
complex farm businesses have developed – both owner occupied
and tenanted from institutional landlords – while farms
rented from private landlords have reduced in number.

During the 1960s in particular the rate of dissolution
of landed estates increased due to the increased tax payable
on classified unearned income, the costs of maintenance
and the inflation of land prices. A series of estates in
North Humberside were sold off either to existing tenants
or to non-local institutional purchasers. In the latter
case, farms would then be relet to existing tenants or
farmed by their new owners with the installation of farm
managers. Such a change in landownership caused a rationalisation of the farm structure throughout the former
estate parishes. In one instance, on a 10 000 ha estate
sold to a trust, eleven farms were amalgamated into three,
hedges were removed, and effective control of the land
was absorbed under a single business organisation.

Where estates have remained, landlords have tended to
expand their own farming operations taking land 'in hand'
when the opportunity arose to reduce the number of tenancies
offered; while the landlord's farming operations have led
to larger field sizes, amalgamation, and the replacement
of outlying farmsteads by modern buildings and equipment.

As Table 7.4 indicates, the resultant landownership
structure in North Humberside is much more complex than
that exhibited in the nineteenth and early periods of the
twentieth century. Public institutions in the 1970s found
it profitable to invest in agricultural land and many
farmers were all too ready to sell their land under lease-
back arrangements so that further working capital could
be released. Not only is the landownership structure more
complex in terms of the groups competing for land, but
also in relation to the effective methods of control that
owners display over their property. For instance, institutional ownership allows varying systems of farm organisation
to develop, ranging from the appointment of employed and
directly answerable farm managers to lease-back agreements
with large family-organised, farm businesses which may own
land and rent land from a variety of other sources. It is
this latter example which relates to a Marxian interpretation of the significance of landed property in modern
capitalism, where

> Ground rent, capitalised as the interest
> of some imaginary capital, constitutes
> the 'value' of land. What is bought and
> sold is not land, but title to the ground
> rent yielded from it. The money laid out

Table 7.4   Landownership in North Humberside

Type of Landlord	Number	%
Owner-occupiers	66	37.9
Private landlord (farmer)	5	2.9
Private landlord (non-farmer)	15	8.6
Family Trust	4	2.3
Local Estate	46	26.4
Church Commissioners	4	2.4
Charity organisation	4	2.4
Trawler company	6	3.4
China Clay company	2	1.1
Crown Estate	16	9.2
Bank/insurance/pension fund	3	1.7
University	1	0.6
Other Public Institutions	2	1.1
	174	100.0

Source: Author's survey, 1977.

is equivalent to an interest-bearing investment. The buyer acquires a claim upon the fruits of labour. Title to the land becomes in short, a form of 'fictitious capital'. If capital is lent out as money, as land and soil, house etc., then it becomes a commodity as capital, or the commodity put into circulation is capital as capital.
(Grundrisse, quoted in Harvey 1982, p. 367).

Land itself becomes a pure financial element both for the competing landowning groups (particularly the institutions) and for the large, family-run farm businesses. With the increasing heterogeneity of landowning and occupance the social advantages of owning land are played down. As Harvey further argues:

This heterogeneity is hard to reconcile with the idea that landlords constitute 'one of the three great classes in capitalist society'. But if we probe hard within the diversity we can begin to spot a central guiding feature in the behaviour of *all* economic agents regardless of exactly who they are and what their immediate interests dictate: this is the increasing tendency to treat the land as a pure financial asset. Herein lies the clue to both the form and the mechanisms of the transition to the purely capitalistic form of private property in land.
(Harvey 1982, p. 347).

THE FARM BUSINESS ORGANISATION

Despite these trends towards treating the land market as a purely financial market, modern capitalist agriculture has at its core the farm family. The economic aims of capitalism have harmonised well with the social aims of the farm family in retaining one or more sons within the expanding family firm. It has been argued elsewhere (Symes and Marsden 1978; Marsden 1979; 1981) that a major characteristic of capitalist farming is the growth of multiple-structured, farm holdings whereby the expansion of the farm has occurred with the acquisition of other farm units spatially separated from the home farm. From a sample of farmers in North Humberside, for example, 57 per cent of farms were characterised by a more complex business structure than the sole proprietorship (or by a husband and wife partnership), while 56 per cent of farmers interviewed were controlling some form of multiple-organised farm.

Both through ownership and occupance (of tenanted
land) and in terms of developing complex business organisations, the farm family has become a more stalwart
institution in the structure of modern capitalist farming.
If one examines the changes in the structure of the family
itself since the mid-nineteenth century it is apparent
that today's completed family size is much smaller, with
rarely more than two or three offspring, and that the
incidence of social and biological failure is now extremely
rare. It would seem that the demands of modern agriculture
ensure certain family processes. In the North Humberside
sample, only one household comprised unmarried siblings
and in only one other were the consequences of infertile
marriage evident. The modern farm family must prove itself
effective for continuity and complex business organisation.

In terms of providing an effective labour supply too,
the farm family plays a much more significant role.
Instances where the family furnishes no more than one
person of working age now account for only one half of the
families compared with two thirds in 1851. Large comple-
ments of family labour, based on father-son/brother-brother
combinations, and often divided among two or more households are now much more common than in the previous century.

THE ORGANISATION OF LABOUR

Despite the increasing use of family labour, the increase
in farm size and the intensification of production, there
has been a continuing decline in the agricultural labour
force since the nineteenth century. As in other parts of
Britain mechanisation displaced much of the labour previously
needed to work the arable farms of East Yorkshire. In 1977,
for example, the average number of agricultural workers was
some 60 per cent fewer than the combined number of farm
servants and day labourers employed per farm in 1851. When
account is taken of the smaller number of farms, the total
demand for hired labour is seen to have fallen by 80 per
cent.

Nonetheless the capitalist farm system still relies
heavily upon hired labour. The family labour force proves
adequate only in a small minority of cases (16 per cent),
and on the remaining farms the size and structure of the
hired labour force varies according to farm size, enterprise combination and the amount of family labour available.
On more than half the sampled farms, hired workers outnumbered family members.

The nature of the hired workforce has altered radically
since 1851. Gone is the distinction between farm servants,
bound to the employer for six months of the year, and the
day labourer whose opportunities for employment varied
according to the season. Today the regular employees are
residentially separate but not necessarily independent of
the farm on which they work. Newby (1977; 1978) has

already provided an extensive account of the economic and ideological dependence of farm workers in arable-growing farming areas.

Where enterprise combinations have incorporated intensive systems of livestock and crop production, farm businesses have increasingly relied upon technically trained staff who fulfil regular and specific work roles rather than the 'Jack-of-all-trades' image of the farm worker. In the 1950s and 1960s, for example, the fluctuating prices of barley led many large-scale producers to retain cereals for animal feed. As a result concentrations of intensive pig units are to be found in North Humberside particularly in the Holderness area. On such farms the business partners and full-time hired staff have usually been formally trained. Often the farms include a sizeable proportion of tied cottages. As Newby (1977) has argued in the context of Suffolk, a farm-centred community characterises large-scale farm businesses where hired staff reside close to the farm. Workers are organised into specific groups related to the respective enterprises developed, while the foreman and managerial staff have non-local backgrounds. Three tiers of hired workers can be identified on multiple organised farms; firstly the managerial staff often residing in purpose-built accommodation and highly trained; secondly the more locally based farm workers who reside either in tied cottages or village council housing, and thirdly varying amounts of casual labour. In attempting to reduce labour costs to a minimum and by introducing new crops into rotations which demand different harvesting periods (e.g. vining peas, sprouts, potatoes and oil seed rape) casual labour has re-emerged as an important third tier of agricultural employment.

Farmers compete in a series of labour markets in order to sustain an effective hired workforce, with managers and foremen being sought on a national level and the hired full-time staff attracted from the locality. Local knowledge of men searching for work continues to be an important aspect whatever type of business organisation has developed. The local farm worker's son is a strong candidate for employment and the farmer protects a certain informality regarding the casual labour supply. The same women come yearly to harvest potatoes and sprouts and attend the annual harvest supper on its completion.

The organisation of hired labour is dependent upon the types of business organisations that have evolved. While bureaucratic organisation need not relate directly to the actual size of farm (see Newby 1977, p. 303), the ways in which the farmer and his co-partners organise their farm business - in terms of level of multiple structure, number of intensive enterprises, levels of specialisation and responsibility among partners - influences the type of farm work undertaken and the degree of responsibility and autonomy conferred on the hired workforce. It is in the large, multiple-household farms where proportionately less family labour is available, that a more bureaucratic structure is more likely to arise. It is here that the organisation permits not only a more perfect division of function between the family and the hired workforce, but also a greater division of managerial responsibility among family members. For instance, on a single unit farm two or more sons may divide responsibility for individual enterprises (pigs, arable cultivations, cattle etc.). Alternatively, where the family firm embraces several widely separated holdings responsibility will normally be allocated at the level of the individual holding rather than the particular enterprise.

SOCIAL AND ECONOMIC DIVISIONS AND THE MERGING OF CAPITALS

In comparing evidence from the nineteenth century with that of the late twentieth century this paper has demonstrated some of the ways in which capitalism has penetrated farming in relation to changing landownership and occupancy, the emergent strength of the farm family in controlling and organising the operations of production, and the changing structure and organisation of the farm labour force. Not only have such processes influenced the organisation and structure of farming in lowland areas but they have afforded greater advantages to those landowners and farmers who were able to engage the economies of scale both through increasing size and multiple structure and by adopting more intensive systems of production. Any diseconomies of very large size are kept to a minimum by evolving multiple farm units each with their respective enterprises and labour forces.

    As a result, whereas the social divisions inherent in nineteenth century agriculture were located between land-lord, tenant and farm worker, in the late twentieth century and greater social divisions exist between those farm businesses which are sufficiently large enough to benefit from economies of scale and those which are not. In this sense, the large-scale multiple-organised farmer has much more in common with his institutional landlord, or private landlord than he does with the neighbouring farmer who is an owner-occupier with a heavy mortgage commitment. By harmonising the goals of the farm family around those of developing large-scale farm businesses many farmers have come to dominate agricultural production, and as such continually threaten the existence of smaller farms.

    The advantages of larger holdings and the ways they act to the detriment of the small farmers were documented by Kautsky in the last century. He argues that such advantages include:

        a bigger proportion of cultivated
        acreage, economies in the consumption
        of labour power, draught animals, the
        implements of labour, efficient utilisation

of all resources, the possibility of
using machinery, division of labour,
specialist management, superiority in
the market, easier access to credit.
(Kautsky, quoted in Banaji 1976, p. 26).

Such advantages are demonstrated in the organisation of
family and hired labour noted above. These processes have
particular significance for the agricultural sector in
comparison with industry, mainly because of the nature of
land as a fixed resource. The expansion of agricultural
businesses, unlike their industrial counterparts, is
dependent upon the monopolisation of land and the concentration of operations which, by necessity, reduces the number
of farms and makes the disappearance of the small farmer
imminent. As Kautsky continues:

Under capitalism we can distinguish two
broad movements, accumulation and centralisation. The centralisation of capital
signifies, by contrast a reunion of the
different scattered capitals into one
capital. In agriculture the big landowner
cannot generally increase his wealth
except through centralisation, reuniting
several holdings into one. In industry
accumulation proceeds independently of
centralisation; a big capital can form
without suppressing the autonomy of the
lesser enterprises. When the suppression
occurs it is the effect of the formation
of big industrial capital. Accumulation
is here the starting point. On the
contrary, when land is fragmented into
different properties, and where small
ownership prevails, large holdings can
only acquire land by centralising several
smaller ones. The disappearance of the
smaller holdings is thus the precondition
of the formation of the large enterprise.
(Kautsky, quoted in Banaji 1976, p. 30).

The various types of business organisation based on family
ownership and control, represent variations in this centralisation process of modern capitalist agriculture, and
simultaneously the inability of the many small farmers to
compete with the dominant farm-family firms in obtaining
extra farm units.

An important extension of Kautsky's argument lies in
the linkages between agricultural and industrial capital,
and in some cases the simultaneous growth of both types of
capitalism. There is evidence of large-scale wealthy farmers
transferring agriculturally generated capital into other
industrial and commercial enterprises either related or
unrelated to farming. Traditionally, such linkages have
taken place within the boundaries of the family itself.
For example the auctioneering, drainage and butchering

businesses may be passed on along with the farm. In other cases, capital has been generated by the farm and then redirected towards off-the-farm ancillary businesses. This can be viewed as an extension of the centralisation of large-scale farming. In the North Humberside study (1979), 32 per cent of farmers interviewed held some sort of off-the-farm interest (23 per cent of the sole unit controllers and 38 per cent of multiple farmers), ranging from agricultural contracting and drainage concerns to directorships of public companies. In the South Holderness region feed milling and agricultural machinery enterprises initiated in the early 1960s by a group of large multiple-organised farmers now extend beyond North Humberside and have attained public company status. The interlocking directorships involved make it impossible to estimate the magnitude and direction of capital flow between the agricultural and commercial sectors.

In the latter part of the twentieth century, agriculture in lowland Britain is becoming increasingly integrated with other systems of production and, as a result, more responsive to external forces (outside the confines of the cyclic model presented in Figure 7.1). This paper has attempted to illuminate at least some of the resultant landownership and organisational forms.

REFERENCES

Banaji, J., 1976. Summary of selected parts of Kautsky's 'The Agrarian Question', *Economy and Society*, 5, pp. 2-49.

Britton, D.K. and Hill, B., 1975. *Size and Efficiency in Farming*, (Farnborough: Saxon House).

Caird, J., 1851. *English Agriculture in 1850-51*, (London: Longman).

Cochrane, W.W., 1958. *Farm Prices - Myth and Reality*, (Minneapolis: University of Minnesota Press).

Dexter, K., 1977. The impact of technology in the political economy of Agriculture, *Journal of Agricultural Economics*, 28, pp. 211-221.

Donaldson, J.G.S. and Donaldson, F., 1970. *Farming in Britain Today*, (London: Allen & Unwin).

Harvey, D., 1982. *The Limits of Capital*, (Oxford: Basil Blackwell).

Marsden, T.K., 1979, *The Socio-economic Structure of Farming in North Humberside: A Study of the Farm Family in a Capitalist Farming Area*, (Unpublished Ph.D. Thesis, University of Hull).

Marsden, T.K., 1981. The social organisation and development of large-scale farming systems, in *Social Problems in Rural Communities*, (Leicester: Institute of British Geographers, Rural Geography Research Group Monograph).

Newby, H., 1977. *The Deferential Worker,* (London: Allen Lane).

Newby, H., 1978. The rural sociology of advanced capitalist societies, in: H. Newby (ed.) *International Perspectives in Rural Sociology,* (Chichester: Wiley).

Ruston, A., 1936. The Ridings of Yorkshire, in J.P. Maxton, (ed.) *Regional Types of British Agriculture,* (London: Allen & Unwin).

Self, P. and Storing, H., 1962. *The State and the Farmer,* (London: Allen & Unwin).

Strickland, H.E., 1812. *A General View of Agriculture of the East Riding of Yorkshire,* (York: Wilson).

Sturmey, S.G., 1955. Owner Farming in England and Wales, 1900-1950. *Manchester School,* 23, pp. 246-268.

Symes, D. and Marsden T.K., 1978. Recent developments in agriculture on North Humberside, *North Humberside Themes,* (Hull: Institute of British Geographers).

Whetham, E.H. 1978. *The Agrarian History of England and Wales, volume 8, 1914-1939.* (London: Cambridge University Press).

8. Agricultural corporatism and conservation politics

GRAHAM COX & PHILIP LOWE

At a time of relentless gloom and despondency in economic affairs, the agricultural industry has no difficulty in presenting itself as one of Britain's more convincing post-war successes. As the National Farmers' Union puts it in a recent brochure:

> There is much of which we can be proud.
> Our farmers and growers, of yesterday
> and today, are responsible for some of
> our greatest national achievements.
> Farming is truly the Backbone of Britain.
> (NFU 1982).

The brochure, however, is part of a campaign to 'repair damage' between those who live and work in the town, and those who work in the country - indicating that acceptance of such claims outside of the industry can no longer be taken for granted (NFU Press Release 87, 29 June, 1982)

Nevertheless, turn the pages of this lavishly illustrated paean and an impressive array of statistics - with no intrusive effort at special pleading - come tumbling out. The total output of British agriculture is greater than that of the entire UK motor industry and double the combined turnover of British Rail and British Airways. In 1946 it took three and one-third square metres to produce wheat for a standard loaf; today we can get a loaf from one and one-third square metres. On average a cow now produces 8500 pints of milk a year. In 1946 it would have been 4000 pints.

These spectacular increases in yields, moreover, are matched by impressive improvements in labour productivity - an area in which, by common consent, the general experience of British industry has been disappointing. Thus an increase in output per person of 43 per cent between 1976 and 1981 compares with a figure of 3 per cent for manufacturing industry; and over the last twenty years productivity in agriculture has increased by more than 150 per cent. Between 1970 and 1980 Britain's self-sufficiency in all food increased from 47 to 60 per cent, and in the types of food we can grow in our climate, from 59 to 75 per cent. This general improvement in our degree of self-sufficiency has been accompanied, it is claimed, by movements in price and income relativities which are overwhelmingly favourable to the consumer. More than this, it is asserted that a prosperous agricultural industry has produced a thriving countryside, attractive for recreation and leisure and sustaining a diverse wildlife; and that further improvements in farming methods and output can be achieved without harm to the environment.

Increasingly, however, there are those who respond to such claims with a pained scepticism. For them the years of agricultural intensification - fostered by government and realised through the adoption of new technologies, the spread of mechanisation and the consolidation of holdings - have transformed the rural environment in ways which are unacceptable when considered in any terms other than the most narrowly economistic. Conservation groups and others who take an interest in such matters are likely to view the statistics presented in 'The Backbone of Britain' with considerable circumspection (Norton Taylor 1982; Body 1982). So far as productivity is concerned, for instance, the claims advanced have to be seen in the context of processes of concentration whereby the largest 10 per cent of farms have come to account for half of Britain's food production. If productivity is measured solely in terms of the size of the workforce, then a crucial statistic is that the number of full-time agricultural workers declined by 400 000 (70 per cent) between 1950 and 1980. On a massive scale human labour has been replaced by chemicals and machinery, with dramatic consequences for rural communities and the rural environment. Indeed, if social and environmental external-ities are taken into account, the achievements of British agriculture seem much more ambivalent (Bowers and Cheshire 1983). Even on narrower grounds, the efficiency of British agriculture has been questioned by some of those who have analysed the full range of subsidies and inputs it enjoys (Centre for Agricultural Strategy 1980; Body 1982; Traill 1982).

Such dissent aside, though, the important point to note is the relative ease with which agriculture can plausibly present itself as a major, economic, success story. In a public opinion poll conducted by MORI, respondents were asked their views as to which out of 13 objectives farmers were particularly good at: "Providing high quality food" topped the responses, followed by "Making a good level of profits" and "Increasing their production each year" (MORI 1983a). If it is objected that this success has been substantially sponsored and subsidised by the state, then the apologists for agriculture can readily point to other industries where the rewards accruing to state patronage are much less evident. In any case, the public on the whole do not regard farmers as being excessively feather-bedded. According to a second MORI poll, though most people

(81 per cent) are aware that British farmers receive subsidies for the food they produce, only a minority (23 per cent) think that farmers are "well looked after by Government", and a majority (56 per cent) believe that subsidies benefit the consumer as well as the farmer(MORI 1983b).

Even so, farmers have received more state aid - successively from the British Government and the European Community - than any other industry outside the state sector. Indeed, government commitment to a secure and prosperous agriculture has remained "virtually unquestioned" for over forty years (Newby 1979, p. 111). During this period, agricultural interests have developed an exceptionally close relationship with government. Discussing the politics of agriculture, Self and Storing (1962, p. 197) suggested that "in the farmers' political vocabulary, the critical word has become 'security'"; and by the early 1960s, the relationship between the NFU and government was already, in their view, "unique in its range and intensity" (Self and Storing 1962, p. 230).

This relationship is as strong and close today, and, as we shall see, has enabled the farming community to retain a strategic advantage in the framing of legislation even in the face of increasingly strident and articulate criticism. In the rest of the chapter, we shall review how this worked over a major piece of legislation concerned with rural conservation - the Wildlife and Countryside Act 1981. First, however, it is necessary to say something about the political resources that the farming lobby can bring to such conflicts.

AGRICULTURAL CORPORATISM

The recent vogue for invoking concepts of corporatism in discussions of the post-war political economy of Britain (Panitch 1980) has resulted, after more careful consideration, in heavily qualified assessments of the explanatory power of such concepts, but even the most sceptical of commentators are careful to exempt agriculture from their general conclusions. Metcalfe and McQuillan (1979), for instance, comment that "the institutional conditions described in neo-corporatist analyses of industrial policy making do not exist, at a national level, or in most industrial sectors". The overriding impression, they suggest, "is of weak organisations, weakly interrelated". However, they carefully note that "the main exception to this generalisation is agriculture where the NFU has a statutorily based, negotiating relationship with Government". Indeed, this has been described as "the archetypal corporatist relationship in British politics" (Marsh 1983, p. 9).

Schmitter (1981) has defined corporatism

> as a mode of policy formation, in which
> formally designated interest associations

are incorporated within the process of
authoritative decision making. As such,
they are officially recognised by the
state not merely as interest intermediaries
but as co-responsible 'partners' in
governance and social guidance.

To a marked extent, this definition does characterise the
relationship between the NFU and government (Grant 1983).
However, noting this does not settle highly relevent
questions about the power of the farming lobby. Self and
Storing, for example, suggested that "the tendency of the
close concordat has been as much to debilitate the Union as
to hamper the Government". Wilson (1977, p. 166), similarly,
has emphasized that, far from being in a position to dictate
terms, the NFU has in fact been highly dependent upon the
goodwill of successive governments. In more general terms,
he seeks to devalue any explanation of the course of
agricultural policy based on the power of special interests
and emphasizes instead the way in which agricultural
interests have benefitted from the sheer inertia of government. Certainly, it is apparent that there has been very
little desire within government to change farm policy in
any fundamental way. Though this may be indicative of
consensus, it is important to appreciate the ways in which
such a consensus has been engineered and sustained.

During the past decade, various factors have ensured
greater prominence and contention for questions of agricultural policy. These include the excesses of the Common
Agricultural Policy; the shift of the burden for agricultural
support from the tax payer to the consumer which followed
entry to the EEC; rising unease over animal walfare; the
growth of conservation concern over new farming practices
and technologies; and political interest in the welfare of
rural communities. Surely, Grant (1983) is mistaken in
commenting that "what is perhaps most striking about agricultural politics is not that it is influenced by a
favourable or unfavourable public opinion, but that it is
conducted in the *absence* of public opinion". On the contrary,
with the passage of time, Laurence Easterbrook's words from
the *British Farmer,* September 1953 seem increasingly
prescient: "The greatest hope for security for farming lies
not so much in legislation as in a convinced public opinion"
(quoted in Self and Storing 1962, p. 230).

Currently, the most serious challenge to the apparently
privileged position of farmers arises from conservationists.
As the Society for the Responsible Use of Resources in
Agriculture observed in the report of its inaugural meeting
held in April 1983, "Farmers are becoming increasingly aware
that they are under threat to their freedom of action
because of the strength of public opinion about the way
land is being managed" (RURAL 1983, p. 57). If 'security'
was the critical word in the farmer's political vocabulary
in the immediate post-war period, it could be argued that,
of late, 'autonomy' has come to occupy a similarly pivotal
role. Agriculture has been exceptional not only in terms

of the level of support which it has consistently enjoyed but also in terms of the remarkable freedom from controls enjoyed by the recipients of that support. As Grant (1983) comments, "the state penetrates beyond the farm gates, but it does so through the acceptable channels of financial aid and technical advice". If we are correct in seeing a corporatist relationship as significant in accounting for this agricultural 'exceptionalism', it has been no less relevant to the relative ease and success with which agricultural interests have resisted any legislative moves which might compromise their autonomy and control at the point of production. For these reasons attempts by some conservation interests to secure a more detailed legislative framework for land-use management have a significance far beyond the immediate issues of landscape and habitat protection (Buller and Lowe 1982).

In recent years, therefore, the corporate groupings representing the interests of farmers and landowners, arising from their position with respect to the division of labour and the sphere of production, have faced opposition from conservation groups, who, as associations of individuals with common interests, inhabit the very different, competitive sphere of politics (see Cawson 1982 for a discussion of these distinctions). The provenance of the Wildlife and Countryside Bill and the circumstances of its passage through Parliament must be understood in the light of the highly distinctive context of agricultural policy making. As Grant (1983) comments, "one benefit of such a corporatist arrangement from the farmers' viewpoint ... is that legislative control of the activities of the farmers is used only as a last resort". More specifically, such established relationships between agricultural interests and government made available to policy makers a corporatist policy option, built upon existing political and administrative structures, and promising continuity and minimal disturbance to the agricultural policy community.

THE BACKGROUND TO THE CONFLICT BETWEEN AGRICULTURE AND CONSERVATION

As the scale of the post-war revolution in agriculture has become apparent, there has been a marked shift in emphasis in conservationist circles from a focus on urban and industrial pressures as the main threat to landscape and wildlife, to a preoccupation with the destructive effects of changing farming practices and technologies. Via a series of controversies, over a period of about twenty years from the mid-1950s to the mid-1970s, conservationists came to recognise the various environmental implications of agricultural intensification. The appearance of industrialised farm buildings first aroused concern; then the dramatic decline in bird numbers drew attention to the side effects of synthetic pesticides. Other issues followed, including the loss of hedgerows, moorland and heathland reclamation, the

drainage of wetlands and the destruction of Sites of Special
Scientific Interest (SSSIs).

Initially, such controversies were viewed independently,
with conservation groups seeking specific remedies through
isolated campaigns. It is only within the last few years
that a more integrated understanding of agricultural change
has been achieved. The Countryside Commission's *New Agricultural Landscapes* study of 1974 was influential in demonstrating that changes in agriculture could comprehensively
alter the landscape rather than just individual features,
though the Commission was optimistic that with proper guidance modern agriculture could produce a new though equally
attractive countryside. Since then, sceptical opinion has
hardened as evidence has accumulated of a general destruction
of habitats - evidence first marshalled in the 1977 report
of the Nature Conservancy Council on *Nature Conservation and
Agriculture*, which concluded that "all changes due to modern-
isation are harmful to wildlife except for a few species
that are able to adapt to the new simplified habitats".

. The totality of agricultural change and its sweeping
impact on the rural environment have drawn attention to the
failure of existing, protective measures to moderate or
deflect the tide of change. National Parks and Areas of
Outstanding Natural Beauty (AONBs) cover 19 per cent of
England and Wales; and SSSIs 4 per cent. The administration
of these designated areas is in the hands of local planning
authorities but they lack any control over farming and
forestry which occupy 90 per cent of the land so designated.
Though, in principle, any development of land is subject
to their control, the definition of development in the
Planning Acts specifically excludes "the use of any land
for the purposes of agriculture or forestry (including
afforestation), and the use for any of these purposes of
any building occupied together with the land so used".

This means that various planning measures and designations intended to safeguard the countryside are quite
impotent in relation to the forces now acknowledged as
dominant in the creation or destruction of rural landscapes
and habitats. Recognition of these shortcomings has fuelled
demands for fundamental reforms. Growing appreciation of
the dynamics of agricultural change has generated criticism
of the whole system of agricultural support, and pressure
has built up for general powers to regulate the environmental impact of agricultural development.

The farming lobby has, not unnaturally, resisted any
such curbs. The NFU and the CLA have been quite prepared
to discuss specific conservation problems, but have stressed
the need to retain the goodwill and voluntary co-operation
of the farming community if practical remedies are to be
found. Many farmers and landowners are indeed keenly aware
of the claims of conservation, and the CLA has long reflected
this interest. The NFU, in contrast, formerly showed little
interest in matters which, though a suitable avocation for

relatively leisured landowners, seemed marginal to the
concerns of the working farmer. However, since the passage
of the 1968 Countryside Act, when the first sustained attempts
were made to introduce some statutory regulation of agricultural development in environmentally sensitive areas, the
NFU has joined the CLA in emphasising its concern for conservation and has ensured the representation of the farming
viewpoint in conservation debates. Together the CLA and
the NFU have responded to the charges of conservationists
by presenting farmers as stewards of the countryside. While
staunchly resisting any form of planning constraint or
encroachment on a farmer's eligibility for governmant
improvement grants they have pressed for payments and tax
incentives for farmers to pursue conservation objectives
and, in particular, to compensate for any potential income
foregone through farming so as to preserve the landscape
and wildlife of an area.

Two particular initiatives stand out. In 1969 the two
organisations joined with the Ministry of Agriculture, the
Nature Conservancy, the RSPB and the Royal Society for Nature
Conservation in setting up the Farming and Wildlife Advisory Group, (FWAG). Its aim was to bring together agric-
ulturalists and conservationists to promote mutual understanding and co-operation. FWAG's avowed principle is that
wildlife conservation and profitable farming need not be
incompatible and that loss of wildlife habitat through
agricultural intensification can best be ameliorated by
encouraging farmers to modify their practices and by
providing necessary advice. FWAG has proved remarkably
influential, not least because it enjoys the patronage of
MAFF. Indeed the Ministry has encouraged its regional
advisers to co-operate as closely as possible with the
county FWAGs. Other environmental groups, in contrast,
enjoy little influence with MAFF.

The second initiative by the CLA and the NFU was the
publication of *Caring for the Countryside* in 1977. This
joint statement of intent acknowledged that farmers have an
important part to play, not only in the production of food
and timber, but also in conserving scenery and wildlife
and it presented a basic conservation guide for their members'
use. The statement was a response to overtures from the
Countryside Commission and the Nature Conservancy Council
who wished to see an accommodation between conservation and
modern agriculture, and they welcomed it enthusiastically.
The Countryside Review Committee on which both were represented along with senior civil servants from the Department
of the Environment and MAFF, declared that: "The goodwill
inherent in this statement should be one of the cornerstones
of future Government policy for conservation" (CRC 1978).
On the premise that "any new policy must have the broad
support of the farming community", it roundly rejected any
imposition of controls over agriculture, arguing that they
would be cumbersome, costly and unconstructive. Instead,
it called for "a voluntary and flexible policy, based on
advice, encouragement, education and financial inducements".

The argument of controls versus voluntary co-operation came to a head over the issue of moorland reclamation on Exmoor. Attention had first been drawn to the ploughing up of open moorland there in the early 1960s. The Exmoor Society, a branch of the CPRE, had commissioned a study of the extent of change which suggested that the very fabric of the National Park was threatened, and an unsuccessful attempt was made to amend the 1968 Countryside Act during its passage through Parliament to give the National Park Authority powers to control moorland conversion in designated areas. Losses of moorland continued, such that by the mid-1970s a fifth of the open land on Exmoor had been enclosed and improved since its designation as a National Park in 1954.

Eventually, pressure from the CPRE and the Countryside Commission led the Government to appoint Lord Porchester to inquire into the issue in 1977. It was largely to implement his recommendations that the Labour Government introduced its Countryside Bill in 1978. The Bill proposed that Ministers should have the power to designate specific areas of open moor or heath within which National Park Authorities would have been able to make moorland conservation orders to prevent ploughing or other agricultural 'improvements' considered likely to be detrimental.

The NFU and the CLA opposed the Bill. So did the Conservative opposition which rejected any form of compulsion or regulation in favour of reliance on the voluntary co-operation of farmers, arguing that more generous compensation would lead to suitable voluntary agreements. Though the Bill had completed its committee stage when the minority Labour Government fell and had been amended to make management agreements the main instrument of reconciliation, the Conservative Whips would not agree to rush through its remaining stages 'on the nod' as they did with other legislation, so it was lost.

RELATIVE STRENGTHS OF THE LOBBIES

Once in office, the Conservatives introduced their own Bill, which eventually became the Wildlife and Countryside Act 1981. Before examining its passage and contents, it is important to compare the strengths of the main protagonists in the debate over the Bill - the farming and landowning interests and the conservation lobby. Pressure groups seeking to influence legislation bring differing skills and competences to the process of negotiation and vary in their ease of access to the parts of the political system which are important at successive stages of the legislative process. It is crucial to appreciate, therefore, just how close and continuous is the relationship between the NFU, the CLA and relevant Ministries, and the extent to which their involvement with numerous pieces of legislation provides recurrent opportunities to present a coherent philosophy to civil servants and politicians.

Conservation groups do not enjoy this intense and sustained involvement in policy making. They are not of central importance to the effective performance of government or the economy and consequently do not have the close symbiotic relationship with senior civil servants which corporate interest groups enjoy. Most conservation groups have reasonable access to the DoE but the Department does not play a role for environmental interests equivalent to that played by MAFF which promotes agricultural interests and is "committed to the farmers' cause" (Wilson 1977, p. 50; see also Richardson, Jordan and Kimber 1978; and Grant 1983). Partly, this is because it has many other functions, and some of them - promotion of the construction industry and mineral extraction, for instance, - are at odds with the function of environmental protection. It, therefore, deals with many interest groups. Indeed, the NFU has more extensive dealings with the Department than most conservation groups, on matters such as the protection of agricultural land, town and country planning, minerals, pollution control and water management.

In the past, conservation groups have looked to two 'quangos' within the tutelage of the DoE, the Nature Conservancy Council and the Countryside Commission, to act as advocates of the cause of rural conservation within government and there are strong informal links between these agencies and the lobby. But their circumscribed authority limits the usefulness of such links. They have small budgets, limited scope for making policy initiatives, and are politically marginal. Such agencies, it has been argued, "create a kind of phoney 'insider' status for some groups in order to reassure them that they have a sympathetic point of access within the government machine" (Grant 1978). To a considerable extent, in fact, the conservation agencies act as negative filters to the conservation lobby: whilst demands opposed by the agencies are unlikely to be taken seriously by government it does not follow that initiatives supported by them will command government attention. Indeed, despite their status as statutory advisers to government, Ministers and senior civil servants are inclined to regard the conservation agencies as pressure groups whose views should be treated with scepticism and whose involvement in central policy making should be carefully circumscribed (Cripps 1979).

Failure to be closely involved with policy formulation at its pre-public stage often means an uphill campaign at later stages against courses of action to which officials and major interests are committed. Good media and parliamentary relations can only partially compensate for this, though on occasion a combination of parliamentary pressure and public censure has enabled conservation groups to take the offensive against recalcitrant government departments and win concessions (Lowe and Goyder 1983). With the Wildlife and Countryside Bill, however, conservation groups confronted interests with extensive parliamentary contacts and, particularly in the case of the NFU, exceptional lobbying expertise. Of course, at the stage of implementing legislation decisive advantage lies with those interests whose co-operation is vital to its successful administration and, if anything, the position of the farming and landowning community has been enhanced by the provisions of the Wildlife and Countryside Act. In contrast, conservation groups not directly involved in the process of implementation may find it hard to sustain the pressure needed to ensure the full realisation of hard won reforms once the issue has passed from intense media attention and parliamentary scrutiny back into the administrative realm.

THE WILDLIFE AND COUNTRYSIDE BILL

The intention to introduce a Wildlife and Countryside Bill was announced on 20 June 1979. But the previous month, within days of the Conservatives taking office, both the CLA and NFU had separate meetings with Agriculture and Environment Ministers to discuss their legislative plans. Broad agreement on the proposed Bill was reached and from this point through to the enactment of the legislation the Government, the CLA and the NFU remained in essential accord on the philosophy of the Bill and their approach to its more contentious aspects. The Government's proposals were set out in six consultation papers published between August and October 1979. Significantly, they were drafted by civil servants in the DoE's Rural Directorate and not by the Government's statutory advisers, the Countryside Commission and the NCC. Indeed, their views were not formally sought until the public consultation stage - surely an indication of political marginality.

The consultation papers proposed a whole series of minor reforms, but the proposals for conserving natural habitats and open moorland in National Parks attracted greatest attention. They embodied the view that control of farming operations was unnecessary and potentially counter-productive. Conservation objectives should be secured instead through the voluntary co-operation of farmers and landowners, encouraged where necessary by management agreements drafted and financed by conservation agencies. The only elements of compulsion proposed were reserve powers to require landowners or tenants to give 12 months' notice of any intention to convert moor or heath to agricultural land in specified parts of National Parks, or to undertake operations which could be detrimental to the scientific interest of selected SSSIs. In either instance the powers would be activated by Ministers and applied to specific areas. Ministers assured the NFU and the CLA that there was no intention to use the reserve powers for National Parks (a similar reserve power in the 1968 Act providing for six months' notice had neven been used) and that only a few especially important SSSIs would be given this extra safe-guard (a maximum of about 40 was suggested out of a total of some 3500 SSSIs).

Environmental groups responded unfavourably and a few, including the Ramblers' Association and the CPRE, were unreservedly hostile but the impact of their response was blunted by the diversity of their prescriptions and the lack of a concerted approach. The environmental lobby is large and diffuse and this militates against strong central co-ordination However, Lord Melchett the Labour Peer, convened a Wildlife Link Committee which sought to rectify this and co-ordinate the response of nature conservation groups. Legislative safeguards for wildlife habitats were central to their counterproposals since Government proposals to protect a small number of SSSIs seemed inadequate in the face of the threats posed by agricultural intensification and afforestation. Rather than select a few 'super-SSSIs' and by implication downgrade the remainder, voluntary conservation groups urged safeguards for all 3500 sites. Owners of SSSIs, they suggested, should be obliged to notify proposed changes of agricultural practice to give the NCC opportunity to negotiate a management agreement and, where a reasonable agreement could not be reached, the Secretary of State should have powers to make an order preventing harmful change to the site.

The NCC, however, took a different view. Ever since calls for improved safeguards for SSSIs were first made in the early 1960s it had been markedly less enthusiastic than voluntary conservation groups in seeking controls which might overstretch its staff and resources and draw it into confrontation with farmers and landowners. Thus, a few minor reservations aside, it broadly accepted the Government's proposal and expressed itself fully satisfied once assurances had been given that the criteria for special protection would be broadened and the NCC consulted before the selection of any site.

This agreement between the government and the NCC set the latter on a collision course with the voluntary conservation groups. Inevitably much of their lobbying during the public consultation period was directed at shifting the NCC's position before the Bill reached Parliament, as it was unlikely that the government would contemplate a change of mind while still enjoying the backing of its official conservation advisers. In the event, just four days before the Bill's second reading (in December 1980), the NCC was won over, following intense pressure from a combined front of nature conservation groups, acting through the Wildlife Link Committee.

Landscape and amenity interests - with the CPRE, the Council for National Parks and the Ramblers' Association taking the lead- continued the co-ordination they had developed for the Labour Government's Bill. They quickly achieved a common stance, stressing the need for order making powers to protect areas of landscape or wildlife interest and a new system of agricultural grants and subsidies to encourage farm enterprises which might contribute to conservation as well as food production. A package of proposals was agreed with the Countryside Commission but rejected by the Nature Conservancy Council on the grounds that it was too sweeping to be politically practicable. This meant that there was no strategic consensus on the Bill across the conservation lobby.

Whereas conservation groups had not begun to prepare their positions on the Bill until the consultation period, the NFU and the CLA had, during the early summer of 1979, discussed their concerns in private consultation with the MAFF and the DoE. They were, in any case, articulating principles to which they had committed themselves during discussions of other legislation. Nevertheless, they responded to the consultation papers with great care, making many detailed points and proposing specific amendments and, being already in broad agreement with the Government, the consultation period provided an opportunity to marshal arguments for the parliamentary debates.

Since the critics of the Government's proposals had the authoritative backing of the Porchester Report (as well as the support of local government interests, including the Association of County Councils) it was crucial for the NFU, the CLA and the Government to demonstrate that Lord Porchester may have been premature in his judgment that a purely voluntary approach had failed. Indeed, they argued that the voluntary approach was capable of achieving far more than the compulsory powers recommended by his Report. In addition the NFU and the CLA were concerned to safeguard their members' interests in determining the ground rules for conservation payments. November 1979 saw negotiations begin between the Exmoor National Park Authority, the CLA and the NFU to draw up financial guidelines for management agreements. The DoE, MAFF and the Countryside Commission sent observers to these highly significant negotiations and, whereas Porchester had proposed a once and for all compensation for the permanent loss of rights to reclaim equal to the loss in land value, the NFU and CLA insisted that those who voluntarily set aside this option should have the right to choose to be compensated by annual payments related to loss of profit. The guidelines, finally signed on 7 April, 1981 (at the time of the Wildlife and Countryside Bill's second reading in the Commons) in effect treat these two schemes as alternatives by allowing farmers to choose between them.

The success of the Exmoor negotiations was important for the Government; front-bench speakers in the second reading debates in both Houses referred to them as indicating a new spirit of compromise which confounded Porchester's worst fears and vindicated the voluntary approach. The NFU and the CLA, meanwhile, viewed the Exmoor agreement as providing a model for similar agreements elsewhere and in their parliamentary briefings made many references to the Exmoor solution as indicating the soundness of a voluntary approach. That such arguments were effective is indicated by the statutory guidelines for compensation relating to management agreements drawn up following the passage of the

Bill which do, indeed, follow the principles established in the Exmoor agreement.

The passage of the Bill attracted considerable public attention and generated a sustained debate in the media, most of which was very sympathetic to the conservationists' case. Much publicity was given to their evidence indicating widespread destruction of landscapes and habitats; such as an NCC survey which suggested that some 13 per cent of SSSIs suffered damage to their wildlife interest each year. In this atmosphere, a combination of filibuster by the Labour opposition and pressure on ministers from some Conservative MPs and Peers won limited concessions from the Government. The most significant was a requirement that all owners and occupiers of SSSIs give the Nature Conservancy three months' notice of their intention to carry out any potentially damaging operations. This became known as 'reciprocal notification', because a clause had already been inserted in the Bill, mainly under pressure from the NFU and the CLA, requiring the NCC to notify the owners and occupiers of SSSIs of the features it wished to be protected, and to specify the operations that might damage them.

Amenity groups achieved less satisfaction, predictably so given the Conservatives' previous opposition to the landscape protection measures embodied in Labour's Countryside Bill. The CPRE and the CNP brought forward the preliminary findings of the Birmingham University Moorland Study which indicated extensive and continuing losses of open moorland to agricultural reclamation and afforestation - not just on Exmoor but nationally (Parry et al. 1981). Despite being urged to keep options open until these findings could be evaluated, Ministers reiterated their belief in a voluntary approach and used their parliamentary majority to defeat amendments aimed at establishing reserve powers for moorland protection. They did, however, accept the need to monitor moorland change within national parks.

The most contentious change to the Bill during its parliamentary passage followed pressure from the agricultural lobby, and related to changes in the farm capital grant scheme introduced the previous year. In an effort to cut civil service staff, the Government had removed the requirement that farmers should seek prior approval from ADAS to carry out work for which they intended to claim grant. As this also removed the possibility of any official persuasion or advice being brought to bear to safeguard natural features or wildlife threatened by improvement schemes, the change in procedures elicited strong protests from conservation organisations. The Government had responded by requiring farmers wishing to carry out work in National Parks or SSSIs to consult respectively the National Park Authorities or the NCC.

Amendments to the Wildlife and Countryside Bill instigated by the MAFF took these new procedures one step (albeit a big step) further. The new clauses required that where

an agricultural grant was refused on conservation grounds,
the objecting authority (whether the NCC or a National Park
Authority) would have to compensate the farmer. The CLA
and NFU, who had always taken care to emphasise the resource
implications of retaining goodwill, welcomed the measure
as providing the necessary financial safeguards and recomp-
ence for farmers affected by conservation objections.
What alarmed conservationists was that this seemed to give
farmers a legal right to agricultural grant-aid since if
they are denied aid on conservation grounds they must be
compensated for the resulting hypothetical 'losses' from
the meagre budgets of the conservation agencies. (Cox and
Lowe 1983).

THE AFTERMATH OF THE ACT

On 30 October, 1981 the Wildlife and Countryside Act received
the Royal Assent after several hundred hours of parliamentary debate stretching over 11 months. Its fundamental
character represented a success for the astute and carefully
sustained lobbying of the NFU and CLA. Although the
Government had conceded a statutory three month notification
of operations for SSSIs, the CLA, in its Annual Report,
admitted that this constituted but "a relatively minor
infraction of the voluntary principle" (*Country Landowner,*
October 1981, p. 28). As against this minimal setback the
CLA could point with satisfaction to a number of gains made
in Parliament: a legal right for owners to be informed of
the existence of an SSI which would be registrable as a
local landcharge (a reform which conservationists had also
promoted); a statutory duty for owners to be consulted
before new designations are made; an arbitration procedure
for management agreements and compensation for diminution
of the capital value of land in respect of 'super-SSSIs'.
Moreover, the NFU and CLA had successfully resisted any move
towards the extension of planning controls to agricultural
operations and had seen the voluntary arrangements developed
on Exmoor endorsed as a model for dealing with similar
problems elsewhere.

The passage of the Act in no sense marked the end of
political conflict over the legislation. Lobbying and
consultation continued on a number of contentious issues
including a code of practice for owners of SSSIs and the
financial guidelines for compensating farmers denied an
agricultural grant. The NFU and CLA were drawn into drafting
both of these and thus continued to have a tactical advantage
over conservation groups who were not consulted until after
drafts had been published.

The implementation of the Act has already been attended
by considerable controversy. A number of its provisions,
particularly the requirement for owners of SSSIs to give
notice of potentially damaging operations, ensure public
prominence for disputes between agriculture and conservation

as will the preparation and annual update of moorland maps
by National Park Authorities. Similarly, provision for
consultation before new SSSI designations will mean more
publicity for the process of designation itself, as witnessed
most spectacularly by the case of West Sedgemoor. Moreover,
it is hardly surprising that reports of damaging operations
being carried out during the three month consultation
period, both in West Sedgemoor and elsewhere, have already
caused alarm in conservationist circles.

The greatest controversy surrounding the Act so far,
however, has centred on the amount of money available for
its implementation and the related issue of how the conservation agencies are to discharge their new powers and duties.
The recurring fear is that financial stringencies may deter
the agencies from pressing their objections to MAFF-supported
schemes of agricultural intensification. A related concern
was that the NCC's wariness of antagonising local farmers
was leading to considerable delay and even reluctance in the
designation of new SSSIs, particularly in areas where agricultural or forestry development was in prospect. First FOE,
and then the RSPB, threatened the NCC with legal action if
it failed to fulfil its statutory duty of designating land
which met its scientific criteria.

There is no doubt that the implementation of the Act
is being closely monitored and that much is at stake. With
close divisions in the Lords and rigid voting on party lines
in the Commons, the passage of the Act was itself divisive,
giving rise to a new sense of quite fundamental conflict
between agriculture and conservation. In the words of an
editorial in *The Times*, the Act might be "the last chance
for the voluntary principle in agricultural planning, the
last attempt to reconcile the interests of farming and conservation without prohibitions" (15 October, 1981).

With a will born of necessity the CLA and the NFU have
embarked on the task of creating a consensus around the Act.
Concerned to ensure that the stewardship practices of farmers
and landowners more than match the rhetoric of the voluntary
case, the CLA has taken care to emphasise the moral oblig-
ations attaching to ownership of an SSSI and to publicise
the positive conservation action - planting trees, managing
hedges and so forth - undertaken by its members. The NFU,
for its part, has initiated 'the Backbone of Britain'
publicity campaign.

Whilst the CLA, the NFU and the Government are committed
to the success of the Act, environmentalists are sceptical.
Some consider that it is bound to fail to halt the decline
of rural habitats and landscapes and the sooner it is discredited the better. In the words of two leading campaigners
"the Act is a dead end, from which another government will
have to retreat before it can advance by a different route"
(MacEwen and MacEwen 1982). The general politicisation of
countryside issues has had a radicalising effect on the
conservation lobby and individual groups are now presenting

a much more agressive and confrontational style in their lobbying than before the Bill. They are looking to medium term campaigns to achieve some controls over agricultural development and the reform of agricultural policy.

Whereas, in the past, conservation groups have regarded themselves as above or beyond party politics the experience of the Bill's passage has drawn them willy nilly into the party arena. They have maintained the links with MPs and Peers forged during the passage of the Bill, and some have begun to reassess their traditional apolitical stance. FOE and the Ramblers' Association, for example, are working within the Liberal and Labour parties to establish a firm commitment to rural conservation, including planning controls over agriculture and forestry and a restructuring of agricultural grants.

The polarisation of debate has made the position and future of the two conservation agencies particularly problematic. For some years both the Countryside Commission and the NCC have pursued policies based on the notion of a consensus between agriculture and conservation and the Act propels them further in this direction. As the Bill progressed through Parliament, this produced a growing rift between voluntary groups and the conservation agencies as the former grew frustrated at the Government's intransigence and the latter acquiesced in the Government's intention.

Suspicions of the agencies amongst environmentalists has been heightened by the recent trend of political appointments to their governing councils and national advisory committees. A majority of the new appointments made by the Conservatives have farming, forestry or landowning interests, and most of them have served or currently serve in an official capacity with the NFU, the CLA or their Scottish or Welsh equivalents (Lowe 1983). The main intention has been to press the two conservation agencies into line behind the implementation of the Act. A similar trend has occurred amongst the ministerial appointees to National Park Authorities (Brotherton and Lowe 1984).

CONCLUSIONS

With its reliance on the goodwill of the farming and landowning community, the provisions of the Wildlife and Countryside Act can be seen as a corporatist policy option dependent on that community policing its own activities and thereby retaining its autonomy and freedom from controls. Fundamental to the points at issue during the framing and passage of the Act were the property rights of farmers. As we have seen, they have a highly privileged position to defend and considerable political resources, which they were able to mobilise in an effective rearguard action against those who argue that legislative expression should be given to a legitimate public interest in the management of rural land.

Their most important resource is the special and long-established relationship between the farming lobby and government. This relationship partly reflects the power of the NFU and the CLA, but derives ultimately from government commitment to the agricultural sector and a particular mode of state intervention, based on perceptions of its success. The agricultural industry has certainly achieved a degree of security and prosperity beyond the dreams of those who formulated its political strategy in the early post-war years. The outlook of the farming and landowning community now seems to be 'what we have, we hold', and they are particularly jealous of the remarkable autonomy enjoyed by the individual producer. The curbs sought by conservationists therefore pose a fundamental challenge to one of their most cherished privileges - hence the considerable rearguard action mounted by the NFU and the CLA. The relationship between the industry and government made possible a corporatist solution, and the political strength of the agricultural lobby ensured that this option was adopted, even in the face of a critical public opinion.

A clear assessment of the Wildlife and Countryside Act is hard to draw at this stage. The 'countryside' is now a thoroughly politicised issue; the conservation lobby has come of age politically; a media debate with considerable momentum, and frequently damaging to the precepts embodied in the Act, has been effectively sustained; and the Government's unpreparedness to move on many of the issues of greatest concern to the conservation lobby may, in the words of the Director of the CPRE, have "simply stoked the fires of fiercer future controversy" (*The Times,* 6 October 1981).

The effective operation of the Act depends crucially on the co-operation of farmers and landowners. Having argued the 'goodwill' case so forcefully, and having identified themselves so explicitly with the philosophy of the Act, a considerable onus is now placed upon the NFU and the CLA to demonstrate the viability of the Act's provisions for habitat and landscape protection. Much is at stake for these organisations and the implementation of the Act places a considerable strain on the authority of their leadership, both internally and externally.

Panitch (1981) pinpoints the issue of internal authority in his definition of corporatism as "a structure within advanced capitalism which integrates organised socio-economic producer groups through a system of representation and co-operative mutual interaction at the leadership level and mobilisation and social control at the mass level". To ensure that farming practices do not diverge from the stewardship claims of the goodwill case will certainly test the mechanisms for "social control at the mass level" within the farming and landowning communities. As Grant (1983, p. 131). points out corporatist arrangements can only continue to work if the NFU is able to discipline its own members so that they abide by agreements made with government and co-operate with the implementation of policy: "On the whole,"

163

he concludes, "the NFU has done this job well, indeed more effectively than any other industrial organisation". The CLA faces a similar challenge, and, immediately following the passage of the Act, its Council actually considered a proposal that "mavericks be ostracised and be asked to leave the association" (*Country Landowner,* December 1981, p. 28). The general view of the Council, though, was that malefactors should be kept within the association so as to be able to educate them (*Country Landowner,* March 1982, p. 17). However, in July 1983, for the first time in the CLA's 76 year history, the Council did decide to expel a member. He was Hughie Batchelor who had recently been gaoled, amidst much publicity, for felling trees on his land in Kent in flagrant defiance of tree preservation orders and a High Court injunction. The NFU also contemplated expelling him but decided that this would diminish the possibility of influencing his future conduct. The contrasting responses of the two organisations to this particular incident reveal the dilemma they face when confronted by behaviour which may discredit the voluntary approach.

Such questions of the internal and external authority of the farming and landowning lobbies relate to the point made by Hirsch (1977) that there is always a choice to be made between control through compulsion and control through the promulgation of some social ethic. There is, because of this, a relationship of flexible inter-dependence between social solidarity and such moral orders which can provide a framework for self-regulated action. Clearly, however, the framework which a moral order can provide must be seen as a depletable public good, and whatever else, the debates over the Wildlife and Countryside Bill demonstrated that the notions of 'stewardship' so central to the NFU/CLA position have suffered serious depletion recently, in the sense that these organisations are now less able to appeal convincingly to such ideas to regulate their members and legitimate their activities. The Act's provisions do, however, mean that 'stewardship', with its corollary of emphasis on voluntary co-operation and freedom from controls, must still be considered the dominant ideology in the countryside even if it is being increasingly questioned outside the farming and landowning lobby (Rose *et al.* 1976). The CLA has warned its members that their claims that private ownership can be relied upon to maintain the beauty of the countryside will now be tested; and "if the performance of landowners is disappointing, it could do more than anything to shake public confidence in private ownership" (*Country Landowner,* September 1981, p. 9).

The success of the NFU and the CLA in placing the goodwill of the farming and landowning community at the centre of the most important piece of countryside legislation for 32 years suggests that they are, themselves, peculiarly able to call upon the goodwill of government. Now, they must do all they can to ensure the active co-operation of their members in making the Act work, and to convince a sceptical conservation movement, as well as the wider public, that the Act is indeed an appropriate and workable solution to conflicts between agriculture and conservation.

164

REFERENCES

Body, R., 1982. *Agriculture: The Triumph and the Shame*, (London: Maurice Temple Smith).

Bowers, J.K. and Cheshire, P., 1983. *Agriculture, the Countryside and Land Use*, (London: Methuen).

Brotherton, I. and Lowe, P., 1984. Agency or instrument: the role of statutory bodies in rural conservation, *Land Use Policy*, 1(2).

Buller, H. and Lowe, P., 1982. Politics and class in rural preservation, in: M.J. Moseley (ed.) *Power, Planning and People in Rural East Anglia*, (Norwich: Centre of East Anglian Studies), pp. 21-41.

Cawson, A., 1982. *Corporatism and Welfare*, (London: Heinemann).

Centre for Agricultural Strategy, 1980. *The Efficiency of British Agriculture*, (Reading: CAS, Report no. 7).

Countryside Review Committee, 1978. *Food Production in the Countryside*, (London: HMSO).

Cox, G. and Lowe, P., 1982. A battle not the war: the politics of the Wildlife and Countryside Act, in: A. Gilg (ed.) *Countryside Planning Yearbook 4*, (Norwich: Geo Books), pp. 48-76.

Cripps, J., 1979. *The Countryside Commission: Government Agency or Pressure Group*, (London: University College, Town Planning Discussion Paper no. 31).

Grant, W., 1978. *Insider Groups, Outsider Groups and Interest Group Strategies in Britain*, (Coventry: University of Warwick, Department of Politics Working Paper no. 19).

Grant, W., 1983. The National Farmers' Union: the classic case of incorporation, in: D. Marsh (ed.) *Pressure Politics: Interest Groups in Britain*, (London: Junction Books), pp. 129-143.

Hirsch, F., 1977. *Social Limits to Growth*, (London: Routledge).

Lowe, P., 1983. A question of bias: political appointments to the Countryside Commission and the Nature Conservancy Council, *Town and Country Planning*, 52, pp. 132-134.

Lowe, P. and Goyder, J., 1983. *Environmental Groups in Politics*, (London: George Allen & Unwin).

MacEwen, A. and MacEwen, M., 1982. An unprincipled Act? *The Planner*, May/June, pp. 69-71.

Marsh, D., (ed.), 1983. *Pressure Politics: Interest Groups in Britain*, (London: Junction Books).

Metcalfe, L. and McQuillan, W., 1979. Corporatism or industrial democracy, *Political Studies*, 27, pp. 266-282.

MORI, 1983a. Public attitudes to conservation and the use of natural resources. The results of a poll conducted in January 1983 among a representative quota sample of 1991 respondents aged 15+ in Great Britain.

MORI, 1983b. Public attitudes towards farmers. The results of a poll conducted in March 1983 among a representative quota sample of 1797 respondents aged 15+ in England and Wales.

Newby, H., 1979. *Green and Pleasant Land?* (London: Hutchinson).

National Farmers' Union, Information Division, 1982. *Farming: The Backbone of Britain,* (London: NFU).

Norton-Taylor, R., 1982. *Whose Land is it Anyway?* (Wellingborough: Turnstone Press).

Panitch, L., 1980. Recent theorizations of corporatism - reflections on a growth industry, *British Journal of Sociology,* 2, pp. 159-187.

Panitch, L., 1981. Trade unions and the capitalist state, *New Left Review,* 125 (January/February), pp. 21-43.

Parry, M., Bruce, A. and Harkness, C., 1981. The plight of British moorlands, *New Scientist,* 28th May.

Porchester, Lord, 1977. *A Study of Exmoor,* (London: HMSO).

Richardson, J.J., Jordan, A.G. and Kimber, R.H., 1978. Lobbying, administrative reform and policy styles: the case of land drainage, *Political Studies,* 26, pp. 47-64.

Rose, D., Saunders, P., Newby, H. and Bell, D., 1976. Ideologies of property: a case study, *Sociological Review,* 24, pp. 699-731.

RURAL - the Society for Responsible Use of Resources in Agriculture, 1983. *Responsible Agriculture, Report of the Inaugural Conference,* April 1983, Report no. 1.

Schmitter, P., 1981. Interest intermediation and regime governability in contemporary Western Europe and North America, in: S.D. Berger (ed.) *Organising Interests in Western Europe,* (Cambridge: Cambridge University Press), pp. 285-327.

Self, P. and Storing, H., 1962. *The State and the Farmer,* (London: George Allen & Unwin).

Traill, B., 1982. Taxes, investment incentives and the cost of agricultural inputs, *Journal of Agricultural Economics,* 33, pp. 1-12.

Wilson, G., 1977. *Special Interests and Policy Making: Agricultural Policies and Politics in Britain and the USA 1956-70,* (London: Wiley).

9. The politics of rural landownership: institutional investors and ,the Northfield Inquiry

RICHARD MUNTON

In recent years there has been increased interest in questions of agricultural land policy. This can be attributed to a number of matters including changes in the pattern of land occupancy, rising land prices, new capital taxes and an over-due re-examination of the relations between farm size and business efficiency (see, for example, Edwards, 1978; Peters, 1978; 1980; Simpson, 1981; Lund and Hill 1979; Rose et al., 1977). Changes of policy cannot be expected to follow rapidly on the heels of independent research inquiry, whatever the findings, but higher expectations of change might reasonably be aroused by the establishment of an official Committee of Inquiry such as the one recently chaired by Lord Northfield into recent changes in agricultural land tenure in Great Britain (Northfield Committee Report 1979). But is such optimism justified? Can it be assumed that committees of inquiry are set up to aid policy formation? Or, if they are, whether they are likely to make radical proposals?

This paper does not seek to answer the question 'Did the Northfield Committee get it right?' Instead, the intention is to indicate why committees of inquiry are an unlikely means of promoting policy change on contentious issues. The argument will be illustrated by reference to the Northfield Committee's deliberations over the purchase of farmland by the financial institutions (insurance companies, pension funds, property funds, property unit trusts and property bonds).

CONTEXT

In Great Britain there are virtually no restrictions on who may hold an interest in farmland. Owners are neither required to demonstrate their need to own land nor their competence to manage it. Their immense discretion is further augmented by the absence in England and Wales of a land register open to public inspection which records beneficial interests in property; and also by the absence of public agencies able to enforce either land redistribution or farm consolidation when land is sold. Furthermore, farmland is exempt from most planning controls over changes of agricultural use and is accorded preferential capital taxation treatment as a business asset. These privileges of ownership have come under scrutiny following the rise in farmland prices during the 1970s and a decline in opportunities to enter the farming industry for those who are not close relatives of existing tenants or owner-occupiers. Some of these developments have been attributed in part to certain new entrants to the land market, and particularly to the financial institutions.

Under pressure from those farming interests that felt threatened by these developments, the Minister of Agriculture, Fisheries and Food, the Right Hon. John Silkin, MP, established a Committee of Inquiry under the chairmanship of Lord Northfield in September 1977. The Committee's terms of reference were: "To examine recent trends in agricultural land acquisition and occupancy as they affect the structure of the agricultural industry; and to report". Within these broad terms of reference the Committee was expressly enjoined to assess the effects of foreign purchasers and financial institutions on the land market. The Committee's report was expected to take only 6-9 months to prepare but it was not published until July 1979.

At the time of the Committee's establishment the financial institutions were regarded with suspicion and even hostility by sections of the farming industry. They were also being investigated by another Committee of Inquiry charged with reviewing their activities within the economy as a whole (Wilson Committee Report 1980). But to many observers it was their purchase of agricultural land, because of its traditional use as a primary resource by private, family enterprise, that was regarded as a singularly inappropriate form of institutional investment. Yet, in the event, the Northfield Committee neither reported in critical terms on the existing activities of the financial institutions nor recommended restrictions on their future purchasing of land. So why did the Committee adopt this position and why was the Report not followed by an outcry from the farming industry? (see also Munton 1982).

ESTABLISHMENT OF THE COMMITTEE

Departmental Committees of Inquiry are set up by government ministers to enquire into areas of current concern. They represent only one kind of enquiry and should not be confused with Royal Commisions, Select Committees of either House of Parliament or Public Inquiries, although they often work in a broadly similar way (Wraith and Lamb 1971). All function within a political context whatever attempts may be made by the initiating Minister or by the committee to minimize the significance of this. Irrespective of how well the committee collects and presents its evidence and marshals its arguments the impact of its report on policy will be largely determined by the motives that led to the committee's establishment in the first place. Committees are sometimes appointed in the belief that they will improve policy formation. Equally, they can represent convenient devices for delaying or avoiding politically unrewarding decisions (Stanyer 1973). In many instances both objectives may be served. To the Minister the first beneficial effect of the Northfield Committee was for it to deflect from him the pressure being brought by farming interest groups. He was applauded for having set up the first committee this century to enquire comprehensively into aspects of agricultural land occupation. He had responded to those who wanted him to 'do something' about land market trends, for many regarded land prices as being far above their 'agricultural worth' as measured in terms of a capitalisation of current and future expected farming incomes or rents. Indeed it was possible to show that land prices had risen faster than net farm incomes, faster than the Financial Times Share Index and had gained in real value despite a high rate of inflation since the 1960s (Northfield Committee Report 1979, pp. 91-97). The farming community, always sensitive to the activities of outsiders, sought scapegoats; and again it could be shown that the rise in land prices had roughly coincided with the growing scale of land purchase made by the financial institutions (*ibid.*, pp. 122-127, 313-332).

Questions remain, however, as to exactly what the Minister, or those pressing him to act, were seeking from an inquiry. The major interest groups, such as the National Farmers' Union and the Country Landowners' Association, may have been content just to get an inquiry, confident that a committee including several members of the farming industry would be bound to find in their favour. At the very least, they would have an additional channel through which to air their views. As for the Minister, he could reasonably have hoped for improved information on tenure trends, but this he could have obtained from an unpublished study already completed by his own staff and from other academic research (e.g. Gibbs and Harrison 1973; Harrison *et al*. 1977; Munton 1975; 1977). Likewise it is unclear what policy uses the Minister had in mind for the Committee's findings, although the Committee's terms of reference may hold important clues to these. Most notably, the Committee was asked to act speedily, to report rather than to recommend, and to produce a set of options rather than a single view.

THE COMMITTEE

Other studies of committees and commissions indicate the importance of four areas to the outcome of inquiries (Chapman 1973). They are the reasons for setting up the committee, in terms of reference, its membership and its style of working. The terms of reference are, self-evidently, central to a committee's deliberations. The wording can determine the emphasis, direction and limits of the inquiry, and this is established by the Minister and his civil servants before the committee ever meet. In principle and by tradition, the committee is bound by its terms of reference although in this case the Northfield Committee concluded that they were unsatisfactory. The Committee decided that it was not prepared merely 'to report' and to outline options but that it would also make recommendations. By this approach some members sought to tie the Minister's hand arguing that having spent so much time and effort establishing facts and sounding opininon they were entitled to make recommendations. This argument was to be undermined by the Committee's inability to agree on key issues. The Committee also noted that the terms of reference failed specifically to mention the ownership of land, only its occupancy and acquisition. But it would have been illogical for the Committee to have skirted round the more politically sensitive subject of ownership. After all, the landlord's approach to rent levels and farm investment can significantly affect the attitude and farming performance of the tenant, the financial institutions are largely owners and not occupiers, and the rate of growth of owner-occupation is significantly affected by the varying rates of revenue and capital taxation as applied to different kinds of landlord (public, private, charitable, institutional, etc.).

That land ownership remains a contentious issue was also reflected in the Committee's membership. Apart from the chairman, who is a Labour Life Peer, it consisted of five people with farming interests, one chartered surveyor, one property adviser, one retired agricultural journalist, one trade union district organizer and one academic. Officially, each member was appointed because of his expertise and long association with the farming industry. This is not in question, but the Committee's careful balance of interests and the fact that some interest groups talked of 'their man' on the Committee seriously detracts from the official view.

Most committees of inquiry adopt a common style of working. Evidence is sought from those with an interest in the matter, usually in writing in the first instance. Some are then called to give oral evidence and others asked for additional written information. The Northfield Committee employed this procedure but added on-the-spot visits and public meetings. But exactly how evidence is collected and interpreted is also affected by the speed with which the committee is expected to report. In the case of the Northfield Committee this meant that there was no formal research programme and by the time its advisors were appointed a number of key decisions on how to collect the information had been taken. For example, it had been decided to approach the parent bodies of the financial institutions for data and not the investing funds individually.

The data passed to the Committee was, therefore, in an aggregated form and the full range of policies and practices employed by individual funds was obscured.

COMMITTEE FUNCTIONS

The Committee sought to achieve three separate but related functions – fact gathering, opinion seeking and policy forming. These became closely interrelated. Evidence continued to arrive throughout the Committee's life. Nevertheless, most arrived in the first six months, hastily put together by hundreds of organizations and individuals to meet the Committee's apparently restricted timetable. The Committee was overwhelmed by the amount of information it received, and this delayed for some months discussion on policy objectives and thereby concealed the real differences of opinion that existed between its members.

The financial institutions' instinctive reaction to the Inquiry was defensive. Outwardly, they sought to play down the significance of the Northfield Inquiry, arguing that they were doing the farming industry a service by investing in it and by pointing out that agricultural investment represented only a very small part of their activities. At the same time their attitudes were characterised by contradictory positions. On the one hand they were conscious of their public image and wanted to be accepted by the farming industry as being model landlords. On the other they were hidebound by their traditional reticence leading to a reluctance to speak out against their detractors. But most important of all, they were confident of the underlying strength of their position. They became increasingly confident that they could successfully appeal to the Committee's own preferences for mixed tenure and a free market in land, a situation in which they felt sure they would have an important and responsible part to play.

A. Fact gathering

The Committee was established amidst ill-informed debate. Many observers grossly exaggerated the rate of land acquisition and the total holdings of the financial institutions. There was no doubt that the financial institutions had very substantial assets and that these were growing very rapidly during the 1970s. Between 1967 and 1978 the value of their total investments grew in real terms by 40 per cent and their property assets by 240 per cent so that by 1978 their property holdings represented 18.6 per cent of all assets as opposed to 8.5 per cent in 1967. But not only did the financial institutions buy large amounts of property, they also diversified into new areas, including farmland. Since the early 1970s they had been purchasing about 20 000 ha per annum, reaching a peak of 30 000 ha in 1977. Between 1974 and 1978 these acquisitions made up about 7 per cent of the total land sold, again peaking at just over 10 per cent in 1977. Yet by 1978 they had still only acquired 213 052 ha or about 1.2 per cent of all farmland (or 1.9 per cent of the area of crops and grass) in Great Britain. About 75 per cent of this lay in England, 20 per cent in Scotland and the rest in Wales. Within England it was unevenly distributed, most consisting of good arable land in the east and south, but in no county did the proportion rise to more than 7½ per cent of the land in crops and grass. About 20 per cent of the area was farmed in hand (for a more up-to-date picture see Savills - RTP 1982).

The institutions claimed that no more than 2 per cent of their total assets, or 7-10 per cent of their property assets, was ever likely to be held as farmland, and in 1978 the proportions stood at only 0.75 per cent and 4 per cent respectively. Only a small minority of institutions, usually the larger ones, had bought land, although many hundreds more held an indirect interest through their purchase of units from property unit trusts with farmland in their port-folios. Few, if any, institutions had reached the 2 per cent level identified. What could not be clearly established was the rationale for this target figure beyond the fact that it broadly corresponded to the proportion of the Gross National Product attributable to the farming industry. This vagueness, continuing disagreement among the funds themselves as to whether farmland was a competitive invest-ment, and the marginal importance of agriculture to the overall allocation of investment monies increased the uncertainty over the scale of the continuing involvement of these funds in the farmland market. In 1977 the total value of farmland sold was around £410 million, and the financial institutions spent about £50-60 million of this, but the net value of all their property acquisitions in that year amounted to £1235 million or three times the value of *all* the agricultural land transacted.

B. Opinion seeking

Numerous submissions were received, most being strong on assertion but weak on fact. Many major organizations favoured institutional investment, including the Royal Institution of Chartered Surveyors and the Country Land-owners' Association, whilst others argued against any intervention in the land market (a form of negative support). It was claimed that the institutions deliberated carefully over their investments and that through the use of reputable firms of land agents they could be expected to make good landlords. The funds, they noted, provided a new source of capital. They not only invested in farm improvements but also through sale and lease-back released the capital locked-up in the occupier's land. This cash could then be employed to expand the farming business. It could also, of course, be used to buy more land, a villa in Majorca or, more remotely, shares in manufacturing industry.

This favourable view was disputed by the farmers' and agricultural workers' unions and most political parties. They felt that farming should remain based on owner-occupied family farms or, where the land was let, landlords should live locally. This circumspection is not wholly misplaced. In a study completed since the Northfield Report was published Worthington (1980) criticises the financial institutions and their tenants for their general ignorance of countryside matters. He places much of the blame for this on firms of land agents. These, he says, feel obliged to seek maximum or near-maximum profits for their clients, especially as they were instrumental in creating the institutional investment land market in the first place.

The tax advantages of 'corporate bodies that never die', the search for capital gains from investing in land and the inflationary effects on land prices of institutional purchasing also formed major objections. But on the basis of evidence submitted to it the Committee accepted that most institutions bought farmland primarily for its long-term rental revenue, revenue that was expected to rise more rapidly than yields from other investments, and only secondarily for the capital gain that could be realized from selling the land at some future date. By comparison with almost all other forms of investment, farmland has a low initial yield (usually in the order of 2-4 per cent per annum), and unless speculating on short-term land price rises, it must be held long-term for the investment to make financial sense. This led the institutions to claim that they would bring some stability to the pattern of ownership, a stability that had been threatened by the introduction of Capital Transfer Tax in 1974.

C. Policy Formation

Policy formation is a complex process. No one account of how it happens can be regarded as incontrovertible. Many important considerations go unstated and unreported. Certain values and positions are taken for granted. Yet three separate elements of the process can be identified and examined - policy context, issues and recommendations, and reconciliation of members' views.

(i) Policy context - Some members of the Committee believed it possible to treat land tenure problems as a set of purely technical, agricultural matters that could be resolved by technical (marginal rates of taxation, conditions of occupancy) adjustments to the present situation. Detailed policy formation was to them a value-free process. At the same time they appreciated the wider political connotations of the Committee's work. They sought to produce an appraisal acceptable to a wide range of political views and by seeking unanimity to give additional weight to their many compromise proposals (see also Thomas 1973). But they failed to produce a unanimous report, getting the worst of all worlds by being forced to accept an articulate note of dissent from two of their members, whilst the rest of the Committee

signed a weak text containing many statements prefaced by
'some of us think that' or 'a majority of us agree that'.
Given the range of established interests represented on the
Committee this was always a likely outcome once the
Minister's request for a set of options had been rejected.

In the more specific context of agricultural land the
Committee came to endorse four basic principles, the first
three of which were of importance to the financial institutions. It argued for a plural system of land tenure in
which the landlord-tenant system, partnerships, owner-
occupation, and a range of farm sizes and ownership types
could all thrive and co-exist. Associated with this principle
was the need to maintain a free market in land, free in the
sense that no one should be barred from acquiring an interest
in it or that their interest should be limited in extent.
The Committee also laid down two primary policy objectives
for agriculture against which the adequacy of the Report's
tenure recommendations were to be assessed. These were the
continued efficiency and competitiveness of British agriculture, and the retention of personal commitment and
incentive in farming.

(ii) Issues and recommendations - On some aspects of institutional investment the Committee commented unfavourably but
in general concluded that the financial institutions posed
less of a threat to the industry than many had maintained
in their evidence. The area they owned was regarded as
modest and even if current rates of acquisition were to
continue for another 40 years the funds would then only own
6-7 per cent of the present agricultural area. The
Committee neither accepted that the institutions were
especially bad landlords nor that they were responsible for
the general level of land prices in the vacant possession
market. The responsibility for high land prices was placed
squarely on the farming community who bought around 75 per
cent of the total land transacted in 1977 compared to less
then 10 per cent acquired by the financial institutions.

It followed that the Committee did not recommend
restrictions be placed on land purchased by the institutions.
Instead it proposed that their acquisitions be monitored
and a set of guidelines be laid down to temper the practices
of all kinds of landlord. The guidelines, to be drawn up
by the industry, were to identify the social responsibilities
of landlords, the conditions to apply in partnership agreements and the extent of in-hand farming appropriate in
particular cases. But the guidelines were not to have
statutory force and the Minister would merely report annually
to Parliament on their observance. The Committee also
argued that in the best interests of agriculture the institutions should let their land and not farm it themselves,
a reflection of the Committee's wish to retain personal
commitment and incentive in farming. Yet no means of restricting the growth of in-hand farming on institutionally-
owned land was put forward. In effect, these mild criticisms
and modest monitoring proposals can be regarded as endorsing
continued institutional investment in agriculture.

(iii) <u>Reconciliation</u> - At the outset of the Inquiry,
Committee members held a range of views on the financial
institutions but none were unutterably opposed to their
investment in farmland. Most reserved their position. Some
were convinced that high land prices were largely determined
by institutional involvement in the land market and argued
that the 'City' ownership of land was out of tune with
traditional rural values. But as Newby and his colleagues
(1978) have shown, many of those farmers who take this view
are substantial capitalist farmers in their own right holding
broadly similar social and political views to 'City' investors
(see also Massey and Catalano 1978). The real argument was
over who should be allowed to divide up the cake. Lord
Northfield frequently reminded farming audiences of their
own prosperity, partly the result of government support,
and that they must accept that others in the community should
not be excluded from a share in this, if only very indirectly
through the institutional ownership of land. The Committee's
unwillingness to condemn the financial institutions out of
hand at its early public meetings led to accusations from
farmers that the Committee had been fixed in favour of the
institutions. Indeed in a sense it had.

Once the evidence demonstrated that the institutions
neither owned a large percentage of the farmed area nor
were likely to do so in the foreseeable future, the serious
debate was over. Most of the Committee were more concerned
about the *scale* rather than the *principle* of institutional
investment and its opponents had destroyed their credibility
by exaggeration and unsubstantiated assertion. The members'
wish for a free market in land, plural tenure and farming
efficiency - principles that were also being upheld by the
financial institutions, by MAFF and by most farmers - meant
that the case for restricting the institutions' activities
became very thin. Moreover, the Committee's members became
embroiled in what to them were more disputatious matters,
such as taxation and the private landlord. In their search
for ways to slow down the loss of tenanted land many of
them came to accept the financial institutions as the one
possible, and largely acceptable, source of expanding landlordism once it became apparent that they could not all
agree to either a major extension of public ownership or
any really significant financial concessions to the private
landlord.

The Committee as a whole had thus moved from a position
of suspicion or studied neutrality to one of tepid endorse-
ment of the financial institutions, but it had done so
without a complete reconciliation of views and for differing
reasons. One member argued that the data were still too
patchy for firm conclusions to be reached, but the main
report was not re-worded to reflect this view. Instead it
is contained in the note of dissent signed by two other
members. In this note it is argued that the financial
institutions bring few benefits to agriculture and could
even be a disruptive influence. In particular, questions
could be raised about the future stability of institutional

ownership and the willingness of institutional landlords
to retain tenancies, matters of which the financial institutions had made much in their evidence. Will fund managers
not sell the land as it falls vacant, rather than re-let it,
taking advantage of the windfall gain represented by the
vacant possession premium? The premium has fluctuated
between 25 and 50 per cent in recent years. In 1977 and
1978, for example, the financial institutions sold a total
of 13 387 ha in England alone, rather more than can be
easily explained away as portfolio rationalization, of which
69.8 per cent was vacant possession land (MAFF 1979), and
the most recent Savills - RTP Report notes a significant
amount of selling of land in 1982 (Savills - RTP 1983).

CONCLUSION

The main value of the Northfield Committee of Inquiry and
its Report has been to collate information and views on
the acquisition and occupancy of farmland. The Inquiry
highlighted inadequacies in data sources and inconsistencies
in the positions of many of those giving evidence to it.
But although it expressed concern over certain tenure trends,
such as the growth of owner-occupation, it failed to give a
clear lead over policy changes. This lack of direction not
only reflects disagreement about how to achieve objectives
but also a realization on the part of Committee members
that in terms of their basic values the solutions were
often less acceptable to them than the 'problems' they
had identified. In the case of the recommendations relating
to the financial institutions, for example, it proved to be
much more important that fund managers were in agreement
with most of the Committee, MAFF and the most articulate
sections of the farming industry (large farmers) over the
need to retain a free market in land, to ensure the absence
of restrictions on the area any individual could own or
occupy and, among objectives for the farming industry, to
insist that economic efficiency and competitiveness should
remain pre-eminent, than the fact that the investing funds
only owned 1.2 per cent of the agricultural area.

    The probability that the Committee in its Report would
largely endorse the *status quo* was greatly increased by
the inclusion among its members of a range of farming and
landed interests and by its attempt to produce a unanimous
report. The slender chance that policy initiatives would
be taken by government on receipt of the Report were
further reduced by a change in administration from Labour
to Conservative while the Committee was still sitting.
The new government had no direct responsibility for the
Report and no obligation to respond to it. To date
(August 1983), only a few of the recommendations contained
in the minority report relating to increased tax concessions
for private landlords have been acted upon. Moreover, the
views on the Report sent to MAFF by the major interest
groups after it was published re-iterate their original

positions on what became the really contentious issues –
capital taxation, security of tenure and the private landlord - and either ignore or even endorse the Committee's
recommendations on the financial institutions whose presence
in the land market contributed more than any other single
reason to the setting up of the Committee.

REFERENCES

Chapman, R.A. (ed.), 1973. *The Role of Commissions in
Policy Making,* (London: Allen & Unwin).

Edwards, C.J.W., 1978. The effects of changing farm size
upon levels of farm fragmentation: a Somerset case
study, *Journal of Agricultural Economics,* 29,
pp. 143-153.

Gibbs, R.S. and Harrison, A., 1973. *Landownership by Public
and Semi-public Bodies in Great Britain,* (University
of Reading, Department of Agricultural Economics,
Miscellaneous Study no. 56).

Harrison, A., Tranter, R.B. and Gibbs, R.S., 1977.
*Landownership by Public and Semi-Public Institutions
in the UK,* (University of Reading, Centre for
Agricultural Strategy, Paper no. 3).

Lund, P.J. and Hill, P., 1979. Farm size, efficiency and
economies of size, *Journal of Agricultural Economics,*
30, pp. 145-158.

MAFF, 1979. Prices of agricultural land in England, *Stats.*
3/79 and 360/79, (London: Government Statistical
Service).

Massey, D. and Catalano, A. 1978. *Capital and Land:
Landownership by Capital in Great Britain,* (London:
Edward Arnold).

Munton, R.J.C., 1975. The state of the agricultural land
market in England 1971-1973: A survey of auctioneers'
property transactions, *Oxford Agrarian Studies,*
4, pp. 111-130.

Munton, R.J.C., 1977. Financial institutions: their ownership of agricultural land in Great Britain, *Area*
9, pp. 29-37.

Munton, R.J.C., 1982. The Northfield Committee Report and
small farms, in: B.J. Marshall and R.B. Tranter
(eds.) *Smallfarming and the Rural Community,*
(Reading: Centre for Agricultural Strategy,
University of Reading), pp. 18-27.

Newby, H., Bell, C., Rose, D. and Saunders, P., 1978.
Property, Paternalism and Power, (London: Hutchinson).

Northfield Committee Report, 1979. *Report of the Committee
of Inquiry into the Acquisition and Occupancy of
Agricultural Land,* Cmnd. 7599, (London: HMSO).

Peters, G.H., 1978. Land ownership: comment on the current
debate, *Journal of Agricultural Economics,* 30,
pp. 209-212.

Peters, G.H., 1980. Some thoughts on capital taxation, *Journal of Agricultural Economics,* 31, pp. 381-397.

Rose, D., Newby, H., Saunders, P. and Bell, C., 1977. Land tenure and official statistics: a research note, *Journal of Agricultural Economics,* 28, pp. 69-75.

Savills - RTP, 1982. *Agricultural Performance Analysis: First Report,* (London: Savills).

Savills - RTP, 1983. *Agricultural Performance Analysis: The 1982 Analysis,* (London: Savills).

Simpson, I.G., 1981. The Northfield Report: structural policy and agricultural efficiency, *Journal of Agricultural Economics,* 32, pp. 123-133.

Stanyer, J., 1973. The Redcliffe-Maud Royal Commission on Local Government, in: R.A. Chapman (ed.) *The Role of Commissions in Policy Making,* (London: Allen & Unwin), pp. 105-142.

Thomas, N.M., 1973. The Seebohm Committee on Personal Social Services, in: R.A. Chapman (ed.) *The Role of Commissions in Policy Making,* (London: Allen & Unwin), pp. 143-173.

Wilson Committee Report, 1980. *Report of the Committee to Review the Functioning of Financial Institutions,* Cmnd. 7937, (London: HMSO).

Worthington, T., 1980. Land agents and the landscape, *Chartered Surveyor,* 112, pp. 379-381.

Wraith, R.E. and Lamb, G.B., 1971. *Public Inquiries as an Instrument of Government,* (London: Allen & Unwin).

SECTION III

LOCALISM
AND
LOCAL PLANNING

10. The social meanings of localism

MARILYN STRATHERN

People whom we, rather rudely, label as informants employ various devices to put the researcher off. The residents of Elmdon, an Essex village, had two rather specialised strategies to draw on.[1] I recall being told that any information we might want could be found by looking it up in the local library. One was also made sensitive to the question of intrusion – I remember acute agonisings over what one could or could not ask without invading privacy. The two strategies are related. The barrier of privacy demarcates the area over which a person has control, precisely because, like anything else he produces, inform- ation may be alienated – that is, no longer measured in reference to the producer, but commoditised as 'local history' or whatever.

In this chapter I am concerned to trace some rather different kinds of relations – to do with the meanings of localism – although in the long run they too bear on notions of persons defined as having attributes in their possession where the attributes have the character of alienable objects.

THE MEANINGS OF 'LOCALISM'

The various case studies presented in the chapters of this section speak to differently contextualised meanings – that is, to the way particular social groups manipulate the notion of being a local or what a local community is. Since in each case one is dealing with people who identify 'localism' out of interests they share, it should come as no surprise that values shared may be put to different ends, or that meanings themselves may be contestable. One of the rubrics of this discussion is that the idea of localism might also be analytically contestable. What I am asserting in this chapter is that ideas are not simply produced in social contexts but are to be understood also in relation to other ideas: what Elmdoners think about gathering information for depositing in the local library has a bearing on what they think about privacy and the divide between public and private property.

Being told to go and look it up is not simply a statement about library facilities: it is also a statement about
property relations. In the same way, ideas about things
'local' are not simply pieces of social imagery or ideology,
about the social realities of local life (cf. Newby et al.
1978, pp. 279 ff). If we are told there is 'local' feeling
about this or that, or that a particular village has a
strong sense of 'local' identity, we have not necessarily
exhausted the meanings of the idiom in searching for some
corresponding segment or local social grouping on the
ground. For instance, quite spectacularly in the Elmdon
case, the idea of a local community or local families
enables people to talk about mobility - both geographical
and social.

    In many ways of course I am traversing familiar ground.
I am suggesting that we should subject 'localism' - a set
of ideas about the significance of localities - to the
same kind of scrutiny that has attended the analysis of
'community' (cf. Bell and Newby 1971; Pahl 1970) and the
'rural idyll' on the one hand (Newby 1977), and of the
'family' (e.g. Sacks 1975; Harris 1981; Bell and Newby 1976;
Whitehead 1976; 1981) and the 'domestic idyll' on the other
(Smith 1978; Jordanova 1981; Oakley 1974). It is not
accidental that both configurations of ideas should have
come under scrutiny, for images about the naturalness of
the 'home' replicate those about the naturalness of the
'community' (Davidoff et al. 1976). Beautifully illustrated
in these works is the way the image of the natural entity
(cf. Stolcke 1981) works in the interests of certain categories of persons, so that it assumes hegemonic character.
The notion of localism, in fact, cuts across both community
and family - it gives an apparent boundedness to both.
Thus in Elmdon it is the local families that are held to
make up the local community.

TWO STYLES OF ANALYSIS

Anthropologists do not just immerse themselves in other
people's lives; they also work at a distance. Normally
there is a cultural boundary between them and their subject
which leads to a special kind of contextualising of their
observations. Paradoxically, the self-conscious maintenance
of a boundary between observer and subject is in fact
crucial to the flow of ideas between them. Such conceptual
spacing is obviously not restricted to anthropological
practice. Thus Bauman writes:

            It is the availability of ... an interpretive
            scheme (in social science discourse),
            which spans one's own and the alien forms
            of life, rather than the romantic
            'immersion' in the alien subject, which
            enables the sociologist to understand
            the alien form of life as a form,

                 instead of dismissing it as a perverted,
                 or infantile, version of his own.
                 (Bauman 1978, p. 222, emphasis as in original).

Notions such as 'society' or 'culture' are schemes of this
sort, which enable one to describe what others are doing in
a self-contained way. The attendant difficulties - that
one can never actually demarcate a culture or society, that
the observer violates its supposed autonomy - belong to
the strategy. They are produced by the methodological
procedures themselves, and as products of a method cannot
independently criticise it.

        The point has a bearing on studies of 'local communities'.
The idea of a 'local community' is as much an artefact as
the idea of a 'society'. The transient observer defines
an entity to which he does not belong, and which therefore
has a describably coherent, independent life ('form') of
its own. The gravitation of British ethnographers to
relatively small scale communities is of course an attempt
to realise such a coherence in what is experientially
graspable. Such studies are frequently criticised for being
parochial. It may even be pointed out that the observer's
idea of 'community' mystifies social reality (cf. Ennew 1980);
that parish-based studies do not attend to the encapsulating
wider society. Two comments are relevant.

        First, we have no methodogical grounds for preferring
one order of analysis over another: not attending to the
wider society cannot be a criticism of a village-based
study - it merely characterises its limitations. There is
nothing more or less arbitrary about studying a whole region
or country - the fact that a region may geographically
encapsulate its component settlements does not mean that a
regional study somehow supercedes, analytically encompasses
or renders those entities theoretically redundant. Preference can only be determined by the analytical interests of
the moment. The second comment concerns mistaking our
subject of study. I have defended the view elsewhere
(Strathern 1982 b, p. 78) that simply because data is
locally gathered (and all data is in a sense locally
gathered) it need not only refer to local realities (cf.
Ennew 1980, p. 5). This, however, rests on an assumption
about the nature of culture that must be made explicit.

        The necessity to be explicit is forced on one by the
following very obvious consideration. Unlike the inhabitants of exotic places who do not have words for 'society'
or 'community', we are dealing with people whose vocabulary
is coterminous with ours. Elmdoners talk about 'community'
and 'locality', about 'society' and 'class'. I am not sure
that they make a noun out of 'local', but if they do not
apprehend 'localism' as such, they certainly have a highly
developed idea of what does and does not make a 'local'
person. In describing British rural communities, therefore,
we have to make a decision about the status that such
concepts - a part of local vocabulary - are to have in our
analyses. The kind of distancing mechanism that the

observer sets up, his framework, will be crucial.

The first choice is to take *sociological reality* for granted. That is, one assumes that there exist 'communities' of people, occupational 'classes', 'networks' of kin, 'families', and so on. The ultimate analytical task is to apprehend the relationship between these categories, their constitution, the way they impinge on one another. In the course of investigation it may be discovered that what people mean by 'community', 'class' and so on idealises or mystifies the basic social realities, and that one should pay attention to structures of hierarchy or domination and such. Nevertheless what is put in the place of the inform- ant's ideology is the observer's detached understanding of the real nature of these social entities, and of the imaginary quality of how people present their relationship to their real conditions of existence. The notion of a village, for example, may speak to a common base for social life or circumscribe working class horizons. Thus in understanding statements about 'local community' as ideology, observers have struggled to identify what social unit this imagery must describe; or else in discarding the concept of 'community' have resurrected its shadow (e.g. Stacey's 'local social system'). In other words there is an ultimate sense in which folk categories - people's ideas of locality etc. - are taken as descriptions of social units.

An alternative choice is to take *cultural reality* for granted. This framework is based on the assumption that people act within a structure of representations through which they explain their behaviour. It leads to a concern with ideas and values. It also leads to a special reading of sociological issues. Social groupings are regarded as dependent on the construction of value, on the 'negotiation' of identities and boundaries. There is no particular correspondence between folk categories and social groupings - idioms may cross-cut interest groups.

With the focus on ideational constructs, social real- ities appear as shifting frames of reference. Indeed, social reality is problematised. What is a village to the observer is not necessarily a village to its residents. In Elmdon this is true in a double sense. First there is the exogenous divide between middle class residents (inclu- ding farmers) on the one hand, and the landless working class on the other. With reference to the outside world, each category more or less agrees that one comprises 'outsiders' and the other 'locals'. The terminology is agreed upon; many of the connotations of the terms are agreed upon - their reference to life style, extra-village associations and so on. This agreement in turn depends on a particular model of the village; for example, on the notion that the village is made up of a core of 'local' families. But when it comes to allocating particular individuals to these categories, agreement as to the basis of classification may cease. It is not just that people can be ascribed differently, but that different ascriptions are employed. Using the identical labels of 'real villager' and 'stranger', middle class residents may see length of residence, custom, occupation as all determining whether or not a person is a villager, while those who most unambiguously claim the label 'real villager' narrow the qualifications down to exclude the bulk of the population, using a rhetoric of kinship connection that is not widely known. There are thus both agreed upon and not agreed upon definitions. The point is that the *labels* emerge with clarity: they are stable reference points, key operators. But the social alignments supposedly marshalled by them are shifting, unstable. Thus the very strength of the *idea* of a locality is set against ambiguous and overlapping categorisations of who in social terms can reckon themselves as 'locals'. And thus the 'village', so evident with its nucleated houses and sign-posts, cannot match any particular demographic configuration. All that people agree upon is that the village is divided into villagers and non-villagers. In other words it is the social boundaries themselves that are the values which are negotiated.

This alternative approach is the mirror opposite of the first. The idioms or representations used by sets of people to talk about themselves are accorded a prior analytical utility. One then investigates the social groupings and categories that are defined (situationally and contextually) in the way the idioms are employed. In one context a socially relevant distinction may divide those who own their houses from those who do not; in another those who depend on local employment from those who travel to work. The Elmdon distinction between being born in the village and married into it creates a notional divide within the category of farmworkers, which in turn represents certain landless families as 'propertied' entities with extra-village contacts. This same model, employed by farmworking families to set certain claims off from others, is also used by land-owning families to define their exclusive relationship to the village as a whole.

This is the approach which I have selected in relation to the Elmdon material. The assumption that lies behind it is that Elmdon idioms are not exhausted by the Elmdon-ness - they can also, for example, be interpreted as 'English'. It is not just that people in Elmdon do or do not agree as to what makes a villager, but that the terms in which the proposition is discussed cannot be understood only by reference to the characteristics of Elmdon itself. The terms belong to a more general discourse about the nature of class formation in English society.

IMAGES OF MOBILITY

'Localism' conjures up several related images; being rooted in a place; the identity that comes from belonging; bounded social horizons; a sense of antiquity and continuity over time. When Elmdoners think of the 'real' villagers, these

are the terms in which they talk. The village boundary is
salient here. There is a sense in which the pool of
villages within a five mile radius from which spouses
were traditionally drawn is also 'local' in respect of the
world beyond. People with different social horizons will,
of course, intend different things when they use the term
on their own behalf and on behalf of others: but whether
they refer to Elmdon as opposed to neighbouring Heydon or
to East Anglia as opposed to the British Isles, they are
drawing attention to the same thing - the boundedness of
people's horizons.

Each element of this imagery is contextualised by other
ideas. One place presupposes others; rootedness recalls
the rootless; belonging, those who do not; while any
boundary points to what lies beyond; and an abstract stress
on continuity intimates the possibility of break. Attachment
to a place in this English village is constructed in terms
of boundary and belonging. The aphorism, the village belongs
to the villagers and the villagers belong to the village,
sets up a relationship of possession, and within the wider
cultural context those who possess are always divided from
the dispossessed. The idea of local attachment, then, is
constructed *in relation to* the consequences of detachment
(cf. Bell and Newby 1971, p. 52). The bounded local community
suggests also that boundaries can be crossed, persons may
be detached, belonging may be transient. The notion of
localism does not only refer to the value of being local;
it refers also to the value of mobility.

These ideas have practical consequences for investigators. The temptation is to go in search of the stable
inner core of real locals. But most of them vanish under
scrutiny. Indeed I would hazard the generalisation that,
since the formation of class society as we know it, it has
not been the 'village of the mind' that has had such salience
but the 'vanishing village' of the mind. Perhaps there is
a sense in which villages have always been vanishing, cores
disintegrating. Being conscious of the ideational nature
of concepts of stability and mobility should release one
from having to evaluate in these terms demographic evidence
about in-turned communities and migration and population
movement. Whatever other meanings are associated with that
of belonging to a place, and its converse, escaping from
it, we have here a model of class formation - not simply
an ideological gloss on local class relations (or a social
imagery in that sense), but a modelling of how people think
about class itself. In this part of the world at least, it
is important that people can see themselves as both fixed
and mobile.

For Elmdon presents a paradox; a set of residents
claiming that they are the 'real' villagers, and another
set very interested in claiming that they do *not* belong.
The paradox is that the sets overlap: some persons may
claim both, depending on context. Middle class residents
can point to a number of criteria to support the contention

that they do not belong to the locality. Landless farm-
workers and other self-designated working class residents
point to the fact that they do not belong through saying
they belong elsewhere - born in the next village, for
instance. A divide between family membership through birth
and through in-marriage is a significant operator. Here,
one of the functions of drawing the boundaries of locality
close to one - seeing Elmdon as cut off from its neighbours
- is to create a divide that in another sense can be seen
to be crossed. It seems particularly important for some
women, for instance, that they can be seen to have married
in from outside the village - that in this sense they are
not bounded by it.

I hesitate to generalise for the women of Elmdon.
Middle class women by and large will have taken up residence
at the same time as their husbands; they will have come
into the village as a unit, and their foreignness is shared.
On the other hand, a number of working class women, born in
the region around the village, will have moved in independently, or one of their parents may have done. And here
their modelling and perceptions of status may differ from
that of men's. These women are kin-keepers, highly competent
in family matters, keeping kin networks open, and they
rather than their husbands tend to give voice to social
aspirations and to see possibilities for mobility not open
to men. Marriage is crucial for women in opening up choice
and domains of decision making where they are principal
actors. They do not have their own marriage alone to draw
on - but may refer to what marriage meant for their mother
or father, or to the choices their children have made.
Women born in the village are thus also able to trace links
beyond it. There is some evidence that these women see
themselves in key boundary positions. The point is prompted
by Bouquet's (n.d.) analysis of Devon farm households,
where the bounded unit is not the village but the farm,
and the farmer's wife is concerned to draw a line between
the farming household and visiting tourists. In the nature
of the domestic services she offers to each, she differen-
tiates her own spheres of activity.

It is important not to trivialise these aspirations.
They relativise the significance of occupation as a local
determinant of status, for it is by no means clear that
men and women put the same value on particular occupations.
Kinship emerges as a critical source of discrimination:
it is through tracing links outside the immediate family
that women often lay claim to sources of status not otherwise apparent. The idea of a fixed place, a locality, and
of the possibility of moving between localities, are a
complementary symbolic resource. Insistence on attachment
- through one's own life history or that of relatives - to
somewhere outside may crucially support the claim that one
is not to be counted as insider. It seems to matter little
what enduring ties exist with the outside place: the
important proposition is that in contradistinction to one's
present fixture as a resident, mobility can be demonstrated.

NATURAL UNITS

Within feminist anthropology, work on Euro-American cultures
has been particularly concerned with the ideational divide
between domestic and political or public domains (cf. the
two early collections of essays edited by Reiter 1975,
and by Rosaldo and Lamphere 1974; the more recent discussions
in Caplan and Bujra 1978; Rogers 1980; Young *et al*. 1981;
Ardener 1981). These debates are relevant to our understanding of the relationship between localism and class.

One ideological dimension of this divide lies in the
drive to classify certain relationships and units within
our society as 'natural'. One such natural entity is the
individual whom we set against society at large. According
to our folk models, the family has a similar status: in
its serving of biological requirements it is to be distinguished from the creative institutions of a wider culture.
Davidoff, L'Esperance and Newby (1976) draw an explicit
analogy between the idealisation of the 'home' and the
idealisation of the 'village community'. Both turn on
concepts of natural consociation. It is not just that the
village community is seen as less artificial than 'unnatural'
urban life, but that social groupings are seen to arise
naturally out of habits of association; we assume that
people who live in proximity will have things in common,
that frequency of daily interaction is an index of commonality. If we take the 'naturalness' of these assumptions as
part of our data rather than a description of it, then we
are led to ask: to what does the notion of naturalness refer?
Here I am concerned with just one aspect of a possible
answer.

Images of consociation feed the imagery of local
community. In turn, these resonate with the emphasis on
the display of household resources and domestic consumption
(Goody 1976; 1982). The habits of daily life are status
indicators, crystallised in household management. In this
sense, the style of life of a local community is a domestic
style of life writ large. Moreover, as these ideas speak
to class formation, they set up the notion that classes
are 'natural' units; and the corollary of this, that mobility
between units is to be differentiated as individual achievement. The concomitant to these conceptualisations of class
is a notion of personhood that sees status as a kind of
property. A paradigm lies in that of labour, conceived
both as belonging to a person and something that can be
appropriated by another to whom it then belongs. The
alienation of labour presupposes both the natural primacy
of possession (a worker owns his labour power prior to
contract) and the artificial implementation of controls
which shifts possession to another. That possession can
be shifted in this way is a crucial, and in terms of the
world's systems rather bizarre, element in the way we think
about personhood.

What distinguishes English from other status formations
is less the character of stratification itself than the fact

that individuals can be seen to move between strata. We posit that society is simultaneously composed of fixed strata and mobile individuals. Society is thus separated from the persons who compose it, and each entity is given a different value, in the same way that labour is separated from the person who activates it, each again being given a different value. Those conceptualisations produce the distinctive postulate of class society - that achievement can be rewarded through self-mobility, that is, a person may better his own class position.

Thus, this modelling of class relations sets up an *inherent* antithesis between fixed strata and mobile individuals. Persons are seen, at one and the same time, as trapped within or defined by their class position, and, by merit or demerit or by fortune or misfortune, as having the potential for altering their position. The fact that individuals appear to work against the givens of their circumstances or that life-chances are indeed conceptualised as chancy, is *part* of the model of class. Analysis should not take at face value the English emphasis on individuality as constituting natural resistance to social units. Such ideas about nature and culture not only speak to notions of evolution and development, but also put the idea of the individual in a particular relationship to the social units he or she is placed in, such that movement between units is theoretically possible. These issues are pertinent to the connotations of localism. The desire to see the local community as a natural unit with a boundary that individuals can cross, in seeking work or marriage partners, replicates the idea of a class that persons can move into or out of. Women's contacts with other households serve a similar purpose. Class is a 'place' that a person can walk away from. But class is also a place to which an individual can belong. In our folk idioms, a person becomes a member of a particular class through possession, through appropriation - where he belongs is defined in terms of what belongs to him.

UNITS OF ANALYSIS

These issues are relevant to what we take as our subject of study. Let me briefly refer to their empirical source, what the residents of Elmdon make of their village.

People have always moved between villages in the area (Robin 1980); this affects their personal status, but does not create institutionalised links between villages. There are no networks of alliances between villages, any more than movement between classes creates cross-class links. This sustains other villages as 'foreign'. People from these outside places may or may not belong to Elmdon depending on the context. To distinguish themselves from outsiders residing within Elmdon, self-styled 'real' villagers put over the idea of a natural kinship community - everyone

is related to everyone else. But this is a specific evaluation. There is no open community in a sociological sense:
rather, individual households assert their privacy; the
streets are empty; old people pass their days without
visitors. Moreover, among those 'all related', ties through
blood are set off against ties through marriage. Marriage
itself stands for differentiation. In spite of the surface
demography which suggests that like marries like, women see
enormous status choices opening up for them, choices they
may also project back on to the marriage of their parents
(see Strathern 1982). The possibilities of status differentiation thus recede to the social unit of the individual
and his or her self-estimation.

    Here I use 'social unit' in an analytical sense. It
is part of our culture, including the culture of social
science, to see the bounded social unit as definitive. Thus
strata are defined first in occupational terms and it is
conventional wisdom to see the husband's occupation as
determining the standing of his household: thus the family
as a unit is encompassed by the class as a unit. Ask people,
again, to describe the village and they will describe it as
a unit, breached by outsiders coming in to live there.
The extraneous presence of strangers does not dissolve the
boundary itself. Where, as in Elmdon, the village is
identified with particular 'real Elmdon' families, the
village as a unit encompasses these families as units.

    In all these cases the very notion of a 'unit' is an
ideological product. The point has been demonstrated for
the family (e.g. Bell and Newby 1976; Whitehead 1976).
Both class and locality idioms work to encompass the family
as a natural entity, a kind of inner unit encompassed by
outer units. The outer unit is public space, and the
conceptualisation of this space - that there *is* a village
separate from its constituent households - raises all sorts
of questions for research. As far as Elmdon is concerned
there is no public life. This disjunction was an endless
source of consternation for Elmdon residents, especially
middle class newcomers, who were always trying to organise
events in which the village could be seen to be a public
entity.

    But to look only to units is to operationalise only
half an ideology. One can only see the family as class-
defined if attention is restricted to a single occupation.
Commonly this is taken to be the occupation of the male
household head. Women are typically encompassed in this
view. Yet from the view of Elmdon women the unit flies
apart. There is no unit: the labouring family on a weekly
wage ceases to hold together as an idea. Estimation of
standing depends on what is done with the wage, depends on
household management, even on how you were brought up, how
you were trained and thus what the possibilities once were.
It depends on the education of the children, the jobs they
might get. Elmdon wives do not have to define themselves
in terms of the husband's employment; they may prefer to

think of what their father did, or work experiences of their
own. We are in a world of mobility, alleged choice, multiple
ascriptions, and claims to status in which the occupation of
the chief wage earner is only one element among many. Indeed,
to some extent, women see themselves moving between households.
It is the wife rather than the husband who makes the household her area of concern; and it is likely to be she who
keeps up links between households (cf. Reiter 1975; Stivens
1978; 1981). Elmdon women keep up or drop links in part
because such contacts affect their personal self-estimations.
In other words, they perceive of themselves as moving between
units because they measure themselves by the distance. In
these terms, there is no class-encompassed family unit.

Nor of course is there any village-encompassed family
unit. In the case of class ascription men and women in
Elmdon have different interests in different models. In
the case of the idea of the village, I would not want to
distinguish men and women as interest groups, but there is
a distinction created by gender symbolism itself. One can
only see the village as a unit composed of families as units –
the 'real Elmdon' families – if one takes the family as
defined by patronym. As surname sets, families can be
bounded. That image flies apart if one looks instead at
the bilateral tracing of kinship, at ties maintained through
blood and marriage that cross patronymic designations. It
tends to be women who keep count, and it is women of course
who change their names on marriage. Yet Elmdon village
women are often remembered by their father's surname, thereby
sustaining multiple identity. Not encompassed by any one
designation, they can be seen to move between families.
Men have a more encompassing 'family' identity than women,
even as they are more completely classifiable by the work
they do.

As a unit, the village or local community vanishes
under many circumstances. Its residents have different
internal models of its structure, and vary in the extent
to which belonging or not-belonging is for them a source
of identity. There are also asymmetries of power and thus
internal differentiation within classes and between households. Moreover the village vanishes when one looks at
ties individually sustained between places, with people
visiting here and there, passing on job information,
joining potato lifting teams, travelling to another pub,
situations in which occupation or family identity cease to
be effective as encompassing definitions of a person's
status. It is not that one has moved analytically from,
say, parochial interests to regional or areal interests.
The notion of a region or area is simply an enlargement of
the notion of a bounded locality, the smaller unit retaining
its character within the wider scope. To look at internal
differentiation or to look at movement between units is to
alter the sociological focus of enquiry. Yet in some
respects this shift is *culturally* accounted for within a
single ideational structure which gives the 'individual'
a value separate from 'society'.

To specific audiences and with specific intention people
make symbolic statements about the nature of the village
or the importance of being able to escape from it. The
statements are symbolic insofar as they are referring to
another order of reality - they may be about the village;
they may also be about the pressures of powerlessness or
status aspirations. There will, in addition, be an ideational context with people drawing on specific structures of
ideas when they talk about these things. For example, a
distinction between male and female may be used to talk
about the distinction between unit identification and
individual detachment, though the values given to male and
female may not be constant and may even be reversed when
statements drawn from different social contexts are put
together. What remains constant is the structure of
distinction, and the subject of these statements - fixed
strata and mobile individuals.

Up to now I have emphasised how women seem more mobile
than men when they recall the chances that led them to go
into service or the opportunities presented through marriage.
Yet in another breath people may talk about how women are
tied to their families while it is up to the man to go into
the world and make the best of the opportunities there.
These are not random and contradictory associations, but
precise representations of difference.

There is a profitable analogy here for the interpretation
of our own activities as social scientists. Our symbolic
productions, our analyses, have to be context dependent.
Very crudely, the line I have indicated is between taking
sociological reality and cultural reality for granted.
These are two ideational contexts, that is, dependent on
the value given to ideas, and they affect the status of
analytical propositions. They involve different relationships between analysis and material.

Anderson and Sharrock (1982) bring a sociology of
knowledge perspective to bear on their scrutiny of sociological practice. Essentially they demonstrate the necessary
teleology of sociological description: 'actions' are taken
as the units of analysis, of which the ensuing descriptions
have to be 'metaphors'. This leads to a vanishing effect.
In being self-conscious about procedure, practitioners
find their data disappear in the display of its collection
and organisation (cf. Bauman 1978, pp. 223-224).

The vanishing effect is in fact typical of certain
symbol constructions; where elements stand in a metaphorical
relationship to one another, attention to one will 'obviate'
the other. If acts and actions are taken as the explicandum,
then descriptions and analyses of them stand for those acts
in a total way - they *are* the acts in the observer's account.
A description of village life is only a symbol of village
life: it is never intended to substitute for experience, but

it is intended as a substitution for the representation of
that experience. We come to 'know' what it is like to live
in a village from the observer's representation. Such a
description bears a figurative or metaphorical relationship
to the object being described. 'Sociological reality' is
taken for real in the sense that this kind on analysis
traditionally addresses its own social context, and in doing
so substitutes certain cultural representations with its
own. In the course of this figurative strategy, these
latter representations - the observer's descriptions and
analyses - necessarily collapse the data into themselves.

What determines the type of symbol (the character of
the relationship between analysis and data) is the natuıe
of its context, the character of the propositions taken for
granted. Anthropology takes different things for granted,
and sets up a different context for its own representations.

Analyses that take 'cultural reality' for granted will
tend to focus on people's ideas about 'action'. In other
words it will deal initially with a relationship in the
data between description and behaviour. Its own descriptions
draw from these representations - seeing things from the
villagers' point of view, taking folk models as an analytical
starting point, using the conceptual categories of the
culture. If this is anthropological strategy, then its own
representations stand in a metonymic or literal relationship to its material. People's ideas are part of its
data, but these same ideas also inform the framework of
analysis so that they are part of the analysis too. Given
that the starting point is a relationship between ideational
and social elements in the data, much anthropological work
is devoted to spelling out the further conceptual relationship between its analytical constructs and folk categories
under study. In this frame, analytical constructs can never
be substitutes: rather, in preserving their own distinctiveness they merely set up a relationship of contiguity with
these folk categories (cf. Sharrock and Anderson 1982).

If what people say is interpreted simply as part of
what they do, then their representations are unproblematic
in the sense that there is no relationship to be investigated.
The notion of 'ideology' as misrepresentation can be seen
as a sophisticated version of this. When what people say
is interpreted as ideology, their representations are seen
to be addressing a sociological reality for which the
observer will substitute his own.

Taking what people say as problematic, on the other
hand, moves towards the kind of metonymic relationship that
focuses on the distinction between the observer's ideas
and folk ideas. People's behaviour involves ideas about
behaviour, but neither can be collapsed into the other,
because the observer himself draws on those ideas. Thus
part of the data is other people's analyses, and part of
the terms of analysis is this ideational data. One can
take a villager's point of view, see things through a

villager's eyes, without assuming that one has come to know
what it is like being a villager. If one approaches that
kind of knowledge, then it is done *through* villagers' own
representations of what it is like. One rapidly regresses
of course - a different kind of vanishing effect - to the
point where being a villager emerges in some contexts but
not in others, and there is no village to be described that
is not in people's minds for particular purposes.

CONCLUSIONS

The limitations of setting up a relational (literal)
symbolisation between data and analysis has been exposed by
Bourdieu (1977). Anthropologists coming into relationship
with other cultures tend to see those cultures relationally,
that is, as a matter of code and communication, since their
own job is to decode and communicate. This limitation, I
have tried to argue, is an entailment of a particular
symbolic strategy. In starting with distinctions in the
data it is necessarily concerned with the terms of those
distinctions for they also contribute to the terms of
analysis.

    The 'idea' of localism, as it is encapsulated in the
Elmdoners' model of the real village, has meaning in relation
to other ideas. I have emphasised its placement with respect
to a conceptualisation of mobility. Together these two
ideas belong to a construction of class that sees society
composed of fixed strata and mobile individuals. If I keep
coming back to the bizarre nature of this construct it is
because we English take it so for granted. What we take
for granted of course is precisely what constitutes our
exotic 'otherness' in relation to the cultures of the world.

NOTE

[1] Details concerning this study and the debts I owe are to
be found in Strathern 1981; 1982; see also the companion
study, Robin 1980.

REFERENCES

Anderson, R.J. and Sharrock, W.W., 1982. Sociological work: some procedures sociologists use for organising phenomena, *Social Analysis,* 11, pp. 79-93.

Ardener, S., 1981. *Women and Space. Ground Rules and Social Maps,* (London: Croom Helm).

Bauman, Z., 1978. *Hermeneutics and Social Science. Approaches to Understanding,* (London: Hutchinson).

Bell, C. and Newby, H., 1971. *Community Studies,* (London: Allen & Unwin).

Bell, C. and Newby, H., 1976. Husbands and wives: the dynamics of the deferential dialectic, in D. Leonard Barker and S. Allen (eds.) *Dependence and Exploitation in Work and Marriage,* (London: Longman).

Bouquet, M., n.d. *Drawing the line: the domestic organisation of farm tourism.* (Unpublished paper, Department of Agricultural Economics, University of Exeter).

Bourdieu, P., 1977. *Outline of a Theory of Practice,* translated by R. Nice. (Cambridge: Cambridge University Press).

Caplan, P. and Bujra, J.M. (eds.) 1978. *Women United, Women Divided,* (London: Tavistock).

Cohen, A.P. (ed.), 1982. *Belonging. Identity and Social Organisation in British Rural Cultures,* (Manchester: Manchester University Press).

Davidoff, L., L'Esperance, J. and Newby, H., 1976. Landscape with figures: home and community in English society, in: J. Mitchell and A. Oakley (eds.) *The Rights and Wrongs of Women,* (London: Penguin Books).

Ennew, J., 1980. *The Western Isles Today,* (Cambridge: Cambridge University Press).

Goody, J., 1976. *Production and Reproduction,* (Cambridge: Cambridge University Press).

Goody, J., 1982. *Cooking, Cuisine and Class,* (Cambridge: Cambridge University Press)

Harris, O., 1981. Households as natural units, in: K. Young et al. (eds.) *Of Marriage and the Market. Women's Subordination in International Perspective,* (London: CSE Books).

Jordanova, L.J., 1981. The history of the family, in: Cambridge Women's Studies Group, *Women in Society,* (London: Virago).

Newby, H., 1977. *The Deferential Worker,* (London: Allen Lane).

Newby, H., Bell, C., Rose, D. and Saunders, P., 1978. *Property, Paternalism and Power: Class and Control in Rural England,* (London: Hutchinson).

Oakley, A., 1974. *Housewife,* (London: Penguin Books).

Pahl, R.E., 1970. *Patterns of Urban Life,* (London: Longman).

Reiter, R., 1975. Men and women in the south of France: public and private domains, in: R. Reiter (ed.) *Toward an Anthropology of women,* (New York: Monthly Review Press).

Reiter, R., (ed.), 1975. *Toward an Anthropology of Women,* (New York: Monthly Review Press).

Robin, J., 1980. *Elmdon: Continuity and Change in a North-West Essex Village, 1861-1964,* (Cambridge: Cambridge University Press).

Rogers, B., 1980, *The Domestication of Women. Discrimination in Developing Societies,* (London: Tavistock).

Rosaldo, M.Z. and Lamphere, L., 1974. *Woman, Culture and Society,* (Stanford: Stanford University Press).

Sacks, K., 1975. Engels revisited: women, the organization of production, and private property, in: R. Reiter (ed.) *Toward an Anthropology of Women,* (New York: Monthly Review Press).

Sharrock, W.W. and Anderson, R.J., 1982. On the demise of the native: some observations on a proposal for ethnography, *Human Studies,* 5, pp. 119-135.

Smith, P., 1978. Domestic labour and Marx's theory of value, in: A. Kuhn and A-M. Wolpe (eds.) *Feminism and Materialism,* (London: Routledge & Kegan Paul).

Stivens, M., 1978. Women and their kin, in: P. Caplan and J.M. Bujra (eds.) *Women United, Women Divided,* (London: Tavistock).

Stivens, M., 1981. Women, kinship and capitalist development in: K. Young et al. (eds.) *Of Marriage and the Market, Women's Subordination in International Perspective,* (London: CSE Books).

Stolcke, V., 1981. Women's labours: the naturalisation of social inequality and women's subordination, in: K. Young et al. (eds.) *Of Marriage and the Market. Women's Subordination in International Perspective,* (London: CSE Books).

Strathern, M., 1982a. The place of kinship: kin, class and village status in Elmdon, Essex in: A.P. Cohen (ed.) *Belonging. Identity and Social Organisation in British Rural Cultures,* (Manchester: Manchester University Press).

Strathern, M., 1982b. The village as an idea: constructs of village-ness in Elmdon, Essex, in: A.P. Cohen (ed.) *Belonging. Identity and Social Organisation in British Rural Cultures,* (Manchester: Manchester University Press).

Whitehead, A., 1976. Sexual antagonism in Herefordshire, in: D. Leonard Barker and S. Allen (eds.) *Dependence and Exploitation in Work and Marriage,* (London: Longman).

Whitehead, A., 1981. "I'm hungry, mum": the politics of domestic budgeting, in: K. Young *et al.* (eds.) *Of Marriage and the Market. Women's Subordination in International Perspective,* (London: CSE Books).

Young, K., Wolkowitz, C. and McCullagh, R., 1981. *Of Marriage and the Market. Women's Subordination in International Perspective,* (London: CSE Books).

11. Women's roles and rural society

SUE STEBBING

In recent years much attention has been focused upon
the relationship between features of the social structure
such as the education and social security systems and the
perpetuation of stereotyped female roles. There have been
few attempts to examine specifically local aspects of this
relationship and, in this country at least, none at all
concerned with female roles in a rural, non-agricultural
context. This paper addresses that issue by taking up the
themes of locality and rurality and showing how these are
in a reciprocal relationship with a traditional perception
of female aptitudes and appropriate behaviour.

In the first section the theoretical basis of the study
is outlined and the research methodology described briefly.
It is suggested that sex roles are the outcome of a socially
constructed definition of reality in which the subjective
and the objective are interdependent, hence role perceptions
and behaviour may vary with locality. The second and third
sections deal with features of a rural settlement which, it
is argued, impose distinctive constraints upon the female
sex role. The second section describes an ideology of
rurality and domesticity which reaffirms a traditional view
of women's place and the third section points to some
objective features of life in small settlements which have
the same effect. However, that social structure is not
deterministic is illustrated by the research finding that
some of the women in the study area were found to have non-
traditional ideal roles and this section concludes by
showing how they maintained their definition of reality
within a social situation which arose from and gave rise to
a conflicting definition.

THE RESEARCH

The theoretical foundation to the study was the phenomenological perspective on the study of roles developed by
Berger and Luckmann (1979). Following Schutz (1964) they
reconcile macro- and micro-levels of societal analysis by
conceptualizing society as both objective and subjective

reality, each reality giving rise to and arising from the other. Roles are seen as the link between society and individual - between objective and subjective reality:

> institutions are embodied in individual experience by means of roles. The roles, objectified linguistically, are an essential ingredient of the objectively available world of any society. By playing roles the individual participates in a social world. By internalizing these roles, the same world becomes subjectively real to him.
> (Berger and Luckmann 1979, p. 91).

The performance of roles is dependent upon the acquisition of knowledge through socialization, but knowledge is itself a socially constructed system of ideas, beliefs, procedural rules and assertions of fact and value (Lengermann, Marconi and Wallace 1978). The child takes the knowledge presented to it and makes the values of society its own. Hence forward this taken-for-granted knowledge informs the individual's assumptions about the reality of the world in which she lives and she behaves in accordance with those assumptions. Thus, reality itself is socially constructed and the action of individuals must be seen in the context of the social system within which socialization occurs. From this perspective an individual's definition of reality is open to change as the result of the acquisition of discrepant knowledge. Therefore the original reality must be continuously reaffirmed by contact with its objective social structures and, most importantly, by interaction with others holding similar definitions of reality. Berger and Luckmann also assert that a single social structure can give rise to more than one perception of its reality since individual knowledge and experience is unique.

Using this theoretical framework it becomes possible to see role perceptions and behaviour not as being determined by the social structure but as arising from it whilst the social structure in turn arises from individual perceptions of reality. It is pertinent to investigate the relationship between individual and society at a local level because the individual's reality-maintenance has a specific spatial location within which the people with whom she interacts routinely are of central importance.

With this as background an extended period of data collection was undertaken in two parishes in East Kent with the aim of investigating the relationship between the female sex role and rural locality. The study area was chosen because it was felt to be fairly typical of much of southern, rural England. It was by no means isolated in terms of distance from towns and main lines of communication; agriculture was no longer a significant employer of labour; there had been substantial post-war housing development; and most of the population travelled to work in one of the nearby towns or even commuted to London.

The parishes contained one large village in which the post-war development had been concentrated, a smaller village, a hamlet and scattered houses and farms. The large village boasted a general store with a sub-post office, a butcher, an ironmonger and a hairdresser. A steadily decreasing bus service passed the edge of the village, linking it with local towns during the daytime. Other parts of the parishes had no shops or services and there was no large scale source of employment within the immediate area. There was no doctor, dentist, ante-natal clinic or chemist. From this brief description it will be apparent that although not a remote rural area the parishes enjoyed very limited shops and services and life was extremely difficult and isolated for those people without a car.

The aim of the fieldwork was to collect data on the role perceptions and behaviour of women living in the study area and to examine these in the context of the local social situation. A variety of techniques was used over a three year period including semi-structured interviews, case studies, participant observation and a questionnaire-based survey of a ten per cent random sample drawn from women on the electoral register (n = 77). The questionnaire was used to collect information on the women's feelings about living in the countryside, their personal histories, especially the employment history, their sex-role perceptions and, in the case of the married respondents, their conjugal-role behaviour. Subsequently, the respondents' 'ideal' sex-roles, that is their beliefs about female aptitudes and appropriate behaviour, were classified as either 'traditional' or 'non-traditional'. The elements of a traditional ideal role were taken to be the belief that a woman's primary role was home centred and nurturant and that she should defer to the greater authority of the male breadwinner. The non-traditionals stressed the equality of the sexes, the importance of self-fulfilment and the desirability of involvement in a job or other activities outside the home. It was found that 82 per cent of the respondents had traditional ideal sex-roles. A final stage of fieldwork followed up six of the fourteen non-traditionals to invest-igate the origins and maintenance of their minority definition of reality.

The next two sections bring together theory and research findings in a discussion of the relationship between the ideal female sex-role, rural ideology and the objective conditions of life in a rural area.

RURAL IDEOLOGY AND THE FEMALE SEX ROLE

The 'rural idyll' is deeply ingrained in English culture (Wiener 1981). The image of the countryside as the location of "a natural way of life: of peace, innocence, and simple virtue" (Williams 1975, p. 9) and of villages as the location of 'community' endures even today in our heavily urbanized society. It is perpetuated in planning and

architecture, through the media, by our literature and poetry
(Williams 1975) and in advertisements ("full of country
goodness").

Davidoff, L'Esperance and Newby (1976) have pointed out
how, in the nineteenth century, this idealization of rural
life became associated with an idealization of home and the
family. At a time when the spread of industrial capitalism
challenged the old order, this ideology served to underpin
the traditional authority of the gentry in the community
and the husband in the home. The woman's role was almost
entirely domestic, at the hub of the family and hence at
the hub of the idealized community as a "still centre" of
calm and goodness: women were the "... lynchpin of the static
community" (Davidoff *et al.* 1976, p. 154).

One finding of this research is interpreted as being a
manifestation of the continuing salience of this ideology.
During the pilot work many of the women interviewed volun-
teered the information that they were 'countrywomen', an
epithet which meant little or nothing to the researcher
and which the women themselves found hard to define.
Questions incorporated into the subsequent survey revealed
that 65 per cent of the sample considered themselves to be
countrywomen and only 27 per cent said that they definitely
were not. The countrywomen could not be distinguished by
class or age but they did tend to have rural backgrounds
and to hold traditional ideal sex roles. Although being a
countrywoman was obviously a source of pride, again these
women found it hard to put into words just what it meant.
However, the following responses between them summarize the
main areas of agreement:

> Rita: someone who is homely and enjoys
> cooking. They are much better homemakers
> than town people. They are more friendly
> and less tired because there are no
> pressures from traffic and neighbours and
> they are more relaxed and genuine because
> they don't have to keep up with the
> neighbours all the time and can be them-
> selves. They are more homespun, that is,
> they hate dressing up.

> Jane: she wears gumboots quite often!
> She likes gardening and animals and tends
> to be more resourceful because everything
> isn't just around the corner. She has
> time to get everything into perspective.

These descriptions surely encapsulate the very essence of
the rural and domestic idylls - the calm, well-ordered home
as a refuge from the whirl of meaningless urban activity,
presided over by a woman in touch with nature and with
life's real values.

The countrywomen appeared to be consciously embracing,
at least mentally, a way of life which they felt was envied

by many town-dwellers. At the same time there was an element
of defensiveness in their assertions of the superiority of
rural life, as if they were sensitive to the possibility
of criticism being levelled at it. They were certainly
aware of flaws in the rural idyll themselves and were liable
to grumble about gossip, lack of services and the dearth
of entertainment, but these drawbacks were emphasized as
being of minor importance compared with the advantages of
community, a healthy environment and a leisurely pace of
life.

A very important function of the countrywoman concept
was that it provided a unifying element in a society within
which, because of its small scale, other divisions were very
visible. Manifest and acknowledged divergence of interests
along the lines of class, length of residence or type of
house (estate/non-estate) would, if stressed, make any claim
to community nonsensical and shatter the illusion of consensus which these women were at pains to emphasize.

The rural and domestic idylls are further reinforced
and reproduced by the Women's Institute which defines
itself as:

> a democratically controlled, educational
> and social organization for countrywomen,
> giving them the opportunity of working
> and learning together to improved the
> quality of life in the community and to
> develop their own skills and talents.
> (National Federation of Women's Institutes 1976, p. 1)

The WI motto is "For Home and Country". Once it was a
radical organization which did much to improve rural conditions and to give rural women confidence and self-respect.
In its evidence to the Wolfenden Committee on voluntary
organizations, however, the NFWI admitted that "its functions
can no longer be considered in any way pioneering". Certainly
in the study area it appeared that the members of the WI
were interested in it only as a social night out once a
month and as somewhere to display or to learn domestic skills.
The members emphasized their friendliness and consensus and
any serious controversy was carefully avoided. Certainly
there was little interest shown in the resolutions which
were to be discussed at the AGM in the Royal Albert Hall
and few members seemed to feel that anything was achieved
by the passing of these resolutions. The women were all
very conscious that anyone they strongly disagreed with at
the WI meeting would subsequently be impossible to avoid
and they therefore preferred to avoid argument.

The WI plays a significant part in the reproduction of
the rural ideology. It gives countrywomen a distinctive
identity which, superficially at least, overrides other
social divisions, it is firmly located in the rural community
and it affirms women's domestic role. Above all, because
it emphasizes consensus it is a powerfully conservative
force perpetuating the dominant view of reality. It must

be stressed that the WI takes the form it does very much at
the insistence of its members even though this might contra-
dict the stated constitution and declared aims of the
organization at a national level. If it reproduces an
ideology which keeps women in a subordinate, domestic role
it is because the women themselves want it that way in
accordance with their perception of the reality of the female
role.

RURAL LIFE AND THE FEMALE SEX ROLE

There are certain structural features of life in a rural
locality which are important in the relationship between
the subjective and objective aspects of the female sex role.
They operate by emphasizing the visibility of the home-
centred, nurturant elements of the female sex role and by
limiting exposure to, and the dissemination of, a conflicting
view. The most important of these are the limited availability of transport and social services and the high visibility
of individuals and the constraints on social interaction
within a small settlement.

Transport is of enormous importance in a rural area
because the lack of it means isolation - isolation from a
full range of shops and services, from employment opportunities, from many sources of entertainment and education, and
from social contact of more than a limited nature. In terms
of this study the net effect is insulation from conflicting
definitions of the reality of the female role.

The public transport service in the study area was
such that it was accessible to a minority of the population,
and even for those it provided a very limited service. It
was perhaps not surprising, therefore, to find that nearly
50 per cent of the women surveyed had their own car and a
further 5 per cent had the unrestricted use of a family
car. The availability of transport, whether public or
private, does not in itself mean unconstrained mobility
of course. Travel costs money, and when resources are
limited choices and decisions have to be made; the decision
about what is an essential journey is an important one in
this context. Furthermore, even having both access to
transport and the money to use it might not be sufficient
to enable a woman to overcome the restrictions of the locality and its dominant ideology. If she perceives her immediate environment as offering all that is needed, she will not
be motivated to go further afield for company, work, entertainment or services. Therefore, the availability of
reasonably priced transport should be seen as an important
enabling mechanism rather than as an end in itself.

The relative lack of social service provision in rural
areas is well documented elsewhere (see, for example,
Walker 1978; Shaw 1979) and the parishes under consideration
fared no better and no worse than many others. For the

purposes of this paper what needs to be pointed out is that where there are no social services provided by the State the burden of unpaid care for the elderly, the sick and the pre-school child usually falls upon women. Women are seen as the caring sex for whom it is little or no hardship to take additional nursing or nurturing responsibilities at home since that is their primary role. A local psychiatric social worker reported that a significant difference between urban and rural referrals to her psychiatric hospital was that many rural referrals were senile women admitted in order to relieve relatives (often daughters) who had no access to daycare or support facilities and who could no longer cope alone.

Within the study area there was also a lack of facilities for the pre-school child. There was no registered child-minder and although there was a playgroup, mothers were expected to stay with their children rather than treat it as a creche. There was a small, private nursery school but the fees put it beyond the means of most parents and it was physically isolated so that only car owners could reach it easily.

Constraints such as these restricted women's freedom of movement in many ways. In particular they made it impossible for some women to get a job and one important research finding was that taking up paid work outside the local area could lead to changes in sex-role perceptions and sex-role behaviour.

The pattern of social interaction amongst the women in the study area was held to be very important because routine mixing with like-minded people plays an important part in the maintenance of an individual's perceptions of reality. It had been established that most of the women in the sample had traditional perceptions of the reality of women's abilities and appropriate behaviour. Restricted opportunities for interaction with women holding non-traditional roles would reinforce the traditional reality rather than introduce a challenge to the stereotype.

It was found that interaction between the women was limited both quantitatively and qualitatively. Firstly, there was simply a lack of choice about the number and type of women with whom an individual had sufficient contact to enable a friendship to develop. Unless she had transport and chose to use it in order to increase her social network a woman was effectively limited to those who lived nearby or with whom she worked. Employment was considered to be an important means by which women might be brought into contact with differing perceptions of reality and, indeed, it proved to have been so for many of the holders of a non-traditional role. However, the majority (58 per cent) of those in paid employment worked locally and there were no local opportunities for jobs involving mixing with a more diverse workforce. Indeed, over 40 per cent worked in their own home or someone else's. Thus, despite the relatively high level of employment (62 per cent of the sample), this was not a means of widening the range of social interaction for most women.

Secondly, an important fact of life in small settlements is that any non-conformity is extremely visible. The researcher found that the respondents placed great emphasis on consensus and played down social divisions such as that between 'locals' and 'newcomers'. However, just keeping oneself to oneself is not necessarily sufficient to avoid suspicion and, as Pahl puts it, those,

> who just want to be left alone and
> maintain a greater degree of independence
> from the local norms and values may be
> seen to present an implicit criticism or
> even a threat to the local social situation.
> (Pahl 1970, p. 2).

If a woman had a non-traditional sex role and behaved accordingly she would find it difficult, if not impossible, to avoid being labelled as different by those around her. Even more importantly, she would be aware of that label; social control operates because an individual imputes expectations to others. Without constant support from people with a similar view of reality it would be very difficult for her to maintain her perceptions and behaviour pattern. It is perhaps significant that the psychiatric social worker previously cited described another category of client as being typical of those from rural areas - the elderly or middle-aged, middle class woman who, though not actually deemed mentally ill, was categorized as eccentric. In other words, her ideas and life-style were not in accordance with those of the people around her and she was too visible to be ignored. It seems reasonable to assume that a good many other potential eccentrics find life in rural areas intolerable and leave to find like-minded companions or anonymity in larger settlements.

The final part of the research was centrally concerned with six of the women who were found to have non-traditional ideal sex roles. In view of what has already been said it is perhaps not surprising that the main conclusion reached about how their perception of reality was maintained was that they serverely limited their local social network. With one exception these women worked and had their close friends outside the parishes. Again with one exception it was found that friends and colleagues at work were most important in confirming the women's ideal sex role, and for four of the five married women husbands provided active and continuous support. It was noteworthy that all but one of the women claimed to be unaware that her view of reality was in any way different from the norm. Just as the holders of a traditional role were effectively isolated from alternative definitions of reality, so the non-traditionals tended to insulate themselves from elements of the local situation which reinforced the conflicting view. By so doing, they were lost as role models to the women whose social networks were locally based.

CONCLUSION

Although it is certainly not suggested that social structure is deterministic of behaviour it has been argued in this chapter that, within a rural locality, there are certain specific ideological and structural constraints upon women's role perceptions. Many of these constraints are, of course, modified by class and especially by income. Nonetheless they operate in a reciprocal relationship with a traditional view of women as being primarily home-centred, nurturant and subordinate to the male breadwinner. Although several specific features have been discussed, overall the effect is of the insulation of women in a rural area from altern- ative definitions of the reality of their role and the reinforcement of the traditional role. The women themselves are partly responsible for this state of affairs. They do not challenge the constraints upon their lives because, by and large, they do not perceive them as such and their everyday lives confirm the reality of the objective social structures. Thus subjective and objective realities are linked in an unbroken chain of cause and effect.

These findings are not of merely academic interest. Implicit in the discussion has been the assumption that the stereotyped female role limits the freedom of women to make real choices about their lives. Many of the constraints upon them in a rural locality could be removed by those responsible for planning and development in the rural regions. In a society allegedly concerned with equity and social justice and which is convinced of the desirability of country life, surely this is a challenge?

REFERENCES

Berger, P. and Luchmann, T., 1979. *The Social Construction of Reality*, (London: Penguin).

Davidoff, L., L'Esperance, J. and Newby, H., 1976. Landscape with figures: home and community in English society, in: J. Mitchell and A. Oakley (eds.) *The Rights and Wrongs of Women*, (London: Penguin).

Lengermann, P., Marconi, K. and Wallace, R., 1978. Sociological theory in teaching sex-roles: Marxism, functionalism and phenomenology, *Women's Studies International Quarterly*, 1, pp. 375-385.

National Federation of Women's Institutes, 1976. *Handbook*, (London: NFWI).

Pahl, R.E., 1970. *Whose City?* (London: Longman).

Schutz, A., 1964. *Studies in Social Theory, Collected Papers*, Vol. II, (The Hague: Martinus Nijhoff).

Shaw, J.M. (ed.), 1979. *Rural Deprivation and Planning*, (Norwich: Geo Books).

Stebbing, S.R., 1982. *Some Aspects of the Relationship Between Rural Social Structure and the Female Sex Role: A Study of Women's Role Behaviour and Role Perceptions in Two Kent Parishes*, (Unpublished Ph.D. Thesis, University of London).

Walker, A. (ed.) 1978. *Rural Poverty: Poverty, Deprivation and Planning in Rural Areas*, (London: Child Poverty Action Group).

Wiener, M.J., 1981. *English Culture and the Decline of the Industrial Spirit, 1850-1980*, (Cambridge: Cambridge University Press).

Williams, R., 1975. *The Country and the City*, (London: Paladin).

12. The social effects of primary school closure

DIANA FORSYTHE

In Britain, plans to close a small rural primary school frequently evoke strong public resistance. Such resistance is based upon the widespread conviction that closing rural schools causes social damage, a belief that has received increasing public expression over the past ten years. The most alarming suggestion that has been made is that closing schools may actually destroy rural communities. This has been asserted in two senses. First, it has been claimed that school closure can pose a threat to the physical survival of settlements: they may depopulate altogether, lose other services because of the loss of the school, or stagnate because planners may not grant development permission to an area without a school. Second, some people fear that school closure can cripple a community in a symbolic sense by causing children to lose their identification with it or by removing a symbol of local identity (Forsythe 1983a).

The language in which such assertions are couched is remarkably similar. In relevant debates in the House of Lords and in *The Times*[1], in planning and education journals, in the arguments of public officials, and in the words of rural citizens, certain phrases and images consistently recur with reference to small rural schools. Among the most frequent are the commonplace that 'the school is the heart of the community', and the corresponding conclusion that school closure 'kills' a community.

Despite apparently widespread acceptance of these beliefs and their frequent use in attempts to influence public policy, an examination of the literature suggests that we know very little about either the social importance of rural schools or the social consequences of closing them. Academics have rarely looked systematically at the long-term consequences of withdrawing rural services. In order to provide information for policy-makers, several government agencies decided in the late 1970s to sponsor two research projects to examine the social, educational, and economic consequences of rural primary school reorganisation. These projects were centred at Aberdeen University and at Aston University. I was project manager of the Scottish research team, and was personally responsible for designing and carrying out research on the social effects of school closure.[2]

My own anthropological training would have led me to study this problem by selecting a small number of study communities scheduled to undergo school closure, and observing these communities closely over a period of years both before and after the closure. However, certain external constraints prevented this approach. The funding agencies wanted comparative data from a wide spectrum of areas and types of communities. Furthermore, before I joined the project it had been agreed that the research would be accomplished in two years and that fieldwork would take place in the Highland and Tayside regions of Scotland. Since neither of these regions was scheduled to close any rural primary schools during the time set out for the study, a straightforward longitudinal study was impossible.

Instead, I designed a synchronic study that relied on comparison to compensate for the missing time dimension. A series of study communities was selected that included schools in all the stages from stability to closure that normally occur through time. In addition to this characteristic, the degree of remoteness of the study communities, the settlement type they represented, and the nature of the local economic base were also varied systematically. In choosing areas that had lost their school, date of closure, enrolment size at closure, and pattern of reorganisation adopted by the local authority were varied as well.

Altogether, 16 different areas were selected for study; these comprised the catchment areas of 16 primary schools that were still open, plus the former catchment areas of 11 schools that had been closed within the past 15 years. Two complementary approaches were taken to the study of these comunities. First, all of them were investigated with an interview programme that focused on how people perceived their local primary school, on whether and how they used the school, and on their evaluation of the school's importance in relation to other services. In addition, parents were asked about their children's experience at the local school and about their own relationship with school and teacher. In areas in which schools had been closed, we inquired about how and why the school had been closed, whether at the time the respondent had anticipated any changes as a result of the closure, and whether any changes had then actually occurred. Finally, respondents whose children had had to change schools because of a closure were asked about this. In all, 457 people were interviewed; half were parents of primary school aged children (Forsythe 1983b).

Second, in several of the study areas I carried out short-term participant observation, living with a local

family and producing case studies on the basis of local-
level fieldwork. Half of these communities had had no recent
school closure; in them the research focused on the nature
of school-community relations. In the other half there had
been a recent closure or closure attempt, and there the
research focused on the local perception of this event.
Additional documentary material was collected from the local
authorities involved in order to illuminate the official
reasons for the closure decisions (Forsythe and Carter 1983).

     Together, these various methods generated considerable
data, both quantitative and qualitative in nature. Although
the project methodology did not produce direct data on the
social effects of primary school closure, it did produce
a great deal of indirect data on the subject. This paper
presents some of these findings.

     In order to understand what is lost to a rural commun-
ity when its school is closed, one must first know something
about the part such schools play in local social life while
they are still in existence. I will therefore begin by
looking at how primary schools are perceived by members of
the surrounding community; at school-community relations;
and at the role of teachers in community leadership. Next,
the range of social functions performed by local primary
schools will be examined. The following section takes up
the problem of closure, examining some of its apparent
effects and relating these effects to the way in which
local authorities go about the closure process. In the
last section, the relationship between local people and those
who make educational policy is discussed.

THE SCHOOL AND THE COMMUNITY

Two concerns prompted me to investigate local people's
perceptions of their school. First, education officials
tend to see schools first and foremost as educational
institutions; however, only a minority of households in
rural Scotland have school-aged children and thus are
direct consumers of the local school's educational function.
That fact, plus the growth of concern with the *social*
functions of rural schools, led me to wonder whether the
general public define the purpose of these schools in the
same way as do policy-makers.

     Second, when officials or experts write about rural
schools, they tend to treat them according to categories
of size (1-teacher schools, 2-teacher schools, etc.),
evaluating them in terms of such scale-related attributes
as amount of peer group contact or types of facilities
available. Policy-makers too often seem to consider rural
schools largely as types. Again, I wondered whether rural
people take the same approach to evaluating their schools.

     Several interview questions were designed to elicit
the respondents' evaluation of the local primary school and

to investigate which factors were important in that judgement. As it turned out, people's judgements were quite consistent in each area, and were clearly based on one specific factor: the identity and the nature of the local teacher(s). The most significant attribute of the teachers from the respondents' standpoint seemed to be neither their personal background nor their extra-curricular social role, but rather whether or not they were perceived to be good teachers, actively involved in giving the children a first-rate academic education. Thus, despite changing ideas in other sectors of society, the respondents in the study maintained a traditional view of schools as purely educational institutions, and of education as mainly concerned with academic achievement. General size-related characteristics mentioned in the literature (such as peer group contact or level of educational facilities) were seen as much less important.

For the small schools we studied, the teachers are virtually the only agents of the school system with whom the community comes into contact. To social scientists, a rural school is an abstract institution; but to local people, the school is the teacher. Not surprisingly, then, relations between school and community usually boil down to relations between the teacher and the community. These depend on the extent to which the teacher and local people (both those with children and those without) are able to interact successfully. A successful blend requires first, some measure of agreement about what should go on within the classroom; second, a shared conception about the way in which teacher and local people ought to relate; and third, enough compatibility of personal style that the teacher is able to communicate with members of the community in a friendly and effective way. Getting the blend right can be difficult, especially in a small community in which a teacher can never be anonymous. For a teacher from a large-scale urban environment, learning the social skills required to adapt successfully to a small face-to-face community may not be easy. As the research showed, the match between teacher and community is not always good.

The case studies revealed quite a range of relations between teachers and local people. All three aspects of the relationship mentioned above were causes of discord. There were complaints from parents about the balance of classroom teaching; disagreements about how much information should flow between teachers and parents; and difficulties in communication caused by personality factors and differences in accustomed social style. On the other hand, results in one or two areas showed that it is possible for teacher-community relations to be strong and mutually supportive on all three levels. What seems always to be true is that this relationship is a delicate one, requiring effort on both sides. Our evidence suggests that success may be more likely if the teacher has had previous experience of living in a small, face-to-face community and if the teacher has some sort of personal link with the area.

Many interview respondents said they would like to see teachers playing a more prominent role in local affairs. But

case study evidence shed some doubt on this: the respondents'
abstract views of what should occur did not correspond
with their reactions when it actually happened. The areas
where the extra-curricular social role adopted by the
teacher was given the strongest support by local people
were those in which the teachers led events which concerned
the school and children, but refrained from any attempt to
exercise leadership in a wider sense. Thus, relations
between school and community may be smoother if a teacher
does not attempt to be a community leader, at least during
the years of active teaching. This conclusion conflicts
somewhat with current interest in having rural teachers
play an active role in community development (Solstad
1981; Murray and MacLeod 1981; see also Lauglo 1982)[3].
However, the evidence of this study is that in Scotland,
at least, the dual role of teacher and development worker
would be difficult to manage.

THE SOCIAL FUNCTIONS OF THE RURAL PRIMARY SCHOOL

As we have seen, rural primary schools seem to be perceived
by members of the local community largely in terms of their
educational function; but when such schools are proposed
for closure, many of the arguments in favour of their
retention centre on their presumed *social* functions.
Discovering what these functions are supposed to be is sometimes difficult: they tend to be implicit in the forebodings
expressed about the consequences of closure. Turning these
assumed consequences around in order to express them in
positive terms, the presence of a local primary school is
most frequently asserted to perform the following social
functions (Forsythe 1983a):

1. Helping to maintain local population levels.
2. Supplying the community with locally-resident
   teachers, who provide useful leadership.
3. Providing a place for local people to meet.
4. Promoting social integration by bringing
   people together.
5. Providing a sense of community identity.

I will consider this list in the light of our Scottish
data.

      The nature of the research did not permit direct study
of the demographic question, but I did look at this question
indirectly. Early in each interview, before identifying
schools as a topic of special interest, we inquired about
the respondent's past and anticipated future migration
decisions and about the reasons for these decisions. Many
people believe that the presence or absence of a local
school affects migration decisions; indeed, a question later
in the interview showed that many of the respondents believed
this. Nevertheless, when asked to explain their own
migration decisions, hardly any of them mentioned factors

that could be called school-related. While this was not a rigorous test of the matter, the data cannot be said to provide support for the first function listed above.

The supposed function of teachers in providing community leadership has already been discussed; the data suggest that this is not an important function of rural schools. Some of the teachers did act as community leaders in matters not concerned with the children, but this tended to cause resentment. More successful leadership was provided in some areas by retired teachers who had remained locally resident. Their active involvement in social affairs did not seem to be resented.

The utility of rural schools as meeting places has been widely mentioned. Many authorities use this as their sole index of a school's social importance. However, again my data suggest that in Scotland, at least, this is not a very important function. One-third (36 per cent) of the respondents did report attending at least one event per year in the local school. But people in most of the areas studied had access to a community hall with much better meeting facilities than are provided by a tiny (and often old) school. Most of these schools were not built as community meeting places and do not really serve as such. Much more important is their function in providing the occasion for social events and social interaction which may or may not take place in the school. I will return to this point below.

The research did provide support for the last two suggested social functions on the list. People in every area clearly saw their local school as an important institution. Shown a list on nine basic services (doctor, church, shop, primary school, pub/hotel, hall, post office, bank and petrol station), two-thirds of the respondents picked the school as being among the three most important services to the community. Every school we looked at provided occasions for social activities enjoyed by local people. These included school concerts, Christmas parties, and outings for the children, plus gatherings to decorate the school building or raise money for the school fund. Sometimes these activities took place in the school itself; in other places, the local hall was used. Both parents and non-parents stressed the importance of the social events that took place in connection with the school.

Schools bring people together at formal events. They also bring people together in other ways. In a small community, the local school and children are a focus of common interest. This function is particularly important for the mothers of young children, for whom the maintenance of social links can be difficult because of the demands of childcare. This problem is exacerbated by the dispersed settlement pattern typical of much of rural Scotland, by the high demographic mobility of the areas studied (Lumb 1980a; Jones 1982), and by the fact that women there often

do not drive. For these isolated women, the school provides
a common bond and a channel of communication. As one of
them suggested, "if there was no school in the glen, the
mothers wouldn't know one another".

Schools also promote social integration by helping
people to co-operate for the common good. In the study
communities, not only mothers but also fathers, other
relatives, and local residents with no children donated time
and materials to help their local school. Although urban
residents often think of rural life as inherently harmonious,
small face-to-face communities are frequently divided by
conflict. Schools are a focus of common interest, and are
likely to be seen as neutral with respect to such conflicts;
thus, they are capable of inspiring people temporarily to
forget their differences and work together.

Because they are a focus of common interest – often
the only such focus within the community – schools can come
to symbolise that interest. In other words, they can act
as a symbol of local community identity. This idea was
expressed by residents in all the case study areas.

> It would be a pity if the school closed
> ... because it would never reopen. If
> the school closes it takes away the
> individuality of the community.

Even in areas where the relationship between teacher and
community is poor, and where parents are uneasy about the
quality of their children's education, the primary school
clearly has social importance.

Thus, quite apart from the question of the educational
value of small primary schools in general, or of any individual school in particular, there is evidence that rural
primary schools do have social value for local residents:
they can perform a number of functions, both material and
symbolic. The relative importance of these functions and
the way in which they are performed varies from case to
case, depending on such factors as the nature of local
social life, the availability of alternative institutions
and meeting places, and the relations between teacher and
community. Over time, the social significance of any
particular school may fluctuate, depending upon local
transport conditions and changes in teaching staff.

In summary, some of the social functions commonly
attributed to rural primary schools do not seem to be very
important, at least in the areas studied. These include
the leadership role of the teacher, the utility of the
school building as a meeting place, and the function of the
school in maintaining population levels. (The latter
requires more careful investigation.) On the other hand,
there is sufficient evidence to assert that primary schools
in remote areas *do* perform three important social functions.
These are: 1. the promotion of social interaction between

people divided by physical and social barriers; 2. the promotion of social integration - motivating people to work
together for what they see as the common good, and thus
creating bonds between the people themselves; and 3. the
symbolic representation of local identity - as one respondent
put it, "a school makes a community".

THE SOCIAL EFFECTS OF CLOSURE

In investigating the social importance of the schools in
the study communities, we asked all the respondents what
would happen if their schools were to disappear: "Would it
make any difference to life in this community if there were
no primary school (here)?" The coding scheme distinguished
between effects on children and effects on the community
in general. About half the respondents said that removing
the school would make a difference to the children; in
almost every case, the difference cited was seen as negative.
The problem most often anticipated was the increase in the
children's travel time to an alternative school, and the
long day such a change would necessitate. (Many of the
children in these areas already travel long distances to
school.)

Only 14 per cent of the respondents said that removing
the school would have *no* effect upon their community.
Among those who did anticipate community-level social
effects, not one expected the change to be neutral or
positive: the anticipated effects were all seen as negative.
Clearly, the respondents contemplated the prospect of
losing their schools with considerable unease. The most
frequently anticipated social effect (22 per cent of the
respondents) was demographic: reduced inmigration, increased
outmigration, and/or changing population distribution within
the area. The next most frequently expected effects were
a generalized feeling of loss - the belief that "something
would be missing" without the school and children nearby
(16 per cent), and the loss of the school as a means of
bringing people together and as a focus for activities
(14 per cent). Only 1 per cent mentioned the loss of the
teacher or the loss of the school building in response to
this question.

The question just discussed was a fictional one; the
schools in question were all open. To look at the actual
experience of closure in retrospect, we inquired of people
in areas that had lost their schools what changes they had
anticipated at the time the school was closed, and then what
changes had actually occurred. Again, the coding distinguished between effects on children and effects on the
community in general. Comparing the reported actual effects
with those that respondents said they had anticipated
showed that for the children, things turned out somewhat
better than anticipated, but that this was not the case for
the communities as a whole.

The proportion of respondents who felt that the children had suffered ill-effects from the closure (13 per cent) was much smaller than the proportion who has expected such effects (34 per cent), whereas the group reporting that the children had actually benefitted from the closure (25 per cent) was quite a lot larger than that which had anticipated good effects for the children (10 per cent). Separating the responses of parents and non-parents[4] showed that parents were more likely to feel that transfer to a larger school had been good for the children (34 per cent of parents gave this response as opposed to only 18 per cent of the non-parents.)

Turning now to the recollected consequences of closure for the local community, 56 per cent of the respondents reporting having anticipated no social effect. (This is much greater than the proportion (14 per cent) who now anticipate no social effect should the current local school be closed.) No one reported having anticipated a positive effect, and 30 per cent had anticipated socially harmful effects. Asked what effects the local closure had actually had, 66 per cent of the respondents reported having observed no harmful social effects; 28 per cent felt that closure had been socially damaging; and only 2 per cent said it had had positive effects for the local community. Thus, the data on the effects of school closure on the community as a whole contrast with those on the effects of closure on the children: unlike the case of child-related effects, there was no significant group of respondents who after having anticipated negative social effects found that closure had actually affected the community in a positive way. In other words, whereas reorganisation seems to have had some unexpected benefits for the children, it is seen as having had virtually no benefits - expected or otherwise - from the standpoint of the communities as a whole. The two types of social damage most frequently reported as having occurred as a result of closure were the loss of a social centre and a loss of community identity.

Comparing the pattern of opinions expressed in communities which had and had not experienced a local closure within the past 15 years suggested two further community-level social effects. First, closure seems to increase the distance between school and community. Even many years after reorganisation, the alternative school remains to both parents and non-parents a more remote institution. This is demonstrated by several of the indices used in the interviews. For example, when asked to pick the three most important out of the list of 9 services, proportionately fewer respondents in former catchment areas (i.e., where there had been a closure) named the primary school as being of either community or personal importance; a much greater proportion of the respondents in former catchment areas answered "don't know" when asked their opinion of the local primary school; and parents whose children were transferred as a result of closure experienced on the average less contact with their children's teacher at the replacement

217

school. Thus, the alternative school does not seem fully
to take the place of one that is closed.

Second, the closure of a primary school against the
wishes of the surrounding community seems to create a
feeling of powerlessness or alienation by demonstrating to
local residents that they have little influence over decisions
about the future of local services. When asked how much
local influence they actually had, people in former catchment
areas were nearly twice as likely to feel that they had
little or no power over such decisions as respondents in
areas that had not experienced recent school closures. This
feeling of powerlessness was associated with considerable
dissatisfaction with the situation as the respondents
perceived it to be. A particular focus of this dissatisfaction was the manner in which the local authority had gone
about the closure.

PUBLIC INVOLVEMENT IN THE CLOSURE PROCESS

The recollections of respondents in former catchment areas
showed that public participation in the closure process is
rarely invited. Indeed, in many cases parents were not even
notified that their children's school was scheduled for
closure: as many parents had learned of impending closures
by gossip or rumour as by notification from the education
authority. Only a minority of local residents - even a
minority of parents - were asked for their views before the
closure. Two-thirds of the respondents in former catchment
areas (and half the parents) felt that consultation before-
hand had been inadequate.

Only a minority of respondents expressed support for
closures that had already taken place in their area.
Comparing the retrospective evaluations of closures that
had taken place between 1967 and 1977 showed that time did
not reduce public opposition to particular instances of
closure. Nor was there any indication that the closure of
very small schools is more acceptable to the public: the
respondents did not share the view of many administrators
that primary schools with tiny enrolments are 'non-viable'.

The case histories of closure show that there is a
link between these various findings: the way an education
authority goes about closing a school is an important factor
in shaping public response to it. Resentment of closure
tends to focus on the *style* of the closure itself. Many
respondents commented that while accepting (with regret)
that there was a valid financial argument for a particular
closure, they had been angered by the lack of consultation
with the community and felt that policy-makers should make
more effort to explain things to the public and to listen
to their views.

There is strong evidence from the case studies that
people are disturbed by the feeling that they have little

or no power over the provision of local services. This was demonstrated clearly in the areas in which primary provision had been reorganised. For example, in one case a school was closed and the children were sent to join the pupils at the neighbouring school. From the records, it appears that the education authority followed the specified closure procedure. However, they closed the school very quickly and made little effort to talk with local people about their anxieties concerning the new arrangements. This closure caused a great deal of resentment; a small group of parents fought the decision and for a time refused to sent their children to the alternative school. This closure has now been accepted, but even today - over 15 years after the event - some local people remain aggrieved about the way in which the education authority closed the school. They feel that a decision was made to remove a vital institution from their community without sufficient public discussion of the matter.

The residents of another study community shared a strikingly different perception concerning the extent of local power. This was an area in which an official decision to close the school had been withdrawn after local people had expressed strong opposition to the closure. Whether this protest caused the proposal to be withdrawn in unclear; analysis of the case suggests that this was not the deciding factor. Nevertheless, in this one area people felt that they did have a realistic chance of influencing planning decisions and pointed to the school as proof of the fact. This was the only case study area in which a significant proportion of the people held this view. In the other areas, people were distinctly pessimistic about the amount of influence they could have, believing that decisions were made far away and simply imposed upon them from above.

The provision of public services to sparsely-populated areas is difficult. The problems involved can easily lead to misunderstanding and conflict between the local authority or central government department responsible for providing a service and the local people who are the clients for that service. From an authority's viewpoint, the provision of services to small numbers of people is expensive: significant financial economies can be expected if services are central-ised to some degree.

For local citizens, on the other hand, the centralis-ation of services is unwelcome. They pay their rates and taxes like people in urban areas, but receive inferior and less accessible services. This causes resentment. There was a tendency for people in all the areas to feel themselves at the mercy of officials in distant regional centres. Though one reference to "cold-blooded, heartless adminis-tration" was an extreme, the feeling was widespread that too much power over the provision of rural services is exercised by urban-based policy-makers who do not have to live with the social consequences of their decisions. This judgement was applied quite explicitly to the provision of educational services.

Those who make such decisions may see this as ungenerous. After all, they too are concerned about the quality of the services provided, and attempts are made to discuss closure with parents in affected areas. A list of formal procedures is followed that requires considerable paperwork for local authorities. Moreover, since education committees make decisions in open meetings which may then be reported in the press before officials can write to affected schools and parents, some of the communication gap between rural citizens and education authorities is an unavoidable consequence of the nature of local government.

Nevertheless, it is clear that rural citizens believe that more could be done to involve them in decisions that will affect their communities. To a considerable degree, this is a question of style: the major cause of the resistance to the school closures I investigated was the *manner* in which power was exercised (or seen as being exercised) by education authorities. People were affronted by the feeling that they were not being consulted in a meaningful way. This research suggests that local authorities should treat public participation as an important part of the planning process rather than as an inconvenient ritual to be performed as rapidly as established procedures allow (Lumb 1980b).

CONCLUSION

People in rural areas worry about school closure. They are uneasy about what will happen to their children and to their communities if local schools are closed. The death imagery often used (i.e., the notion that 'school closure will kill the community') is an expression of that fear. Policy-makers should recognize this fear and should try to reduce it by communicating with members of the public, insuring that to the extent possible, arrangements for school provision meet their needs and desires. There is strong evidence that the attitude of education officials makes a great difference to the way in which people experience a school closure.

However, in reality, closure does not turn out to be as bad as people often fear it will, at least with respect to its social effects. While few people find that closure has positive social consequences, their evaluation of the state of things after the fact is not as negative as the picture feared by people whose schools are under threat. There are several possible reasons for this. They include first, the fact that closure and transfer to a larger, usually newer school sometimes turns out to have unexpected benefits for the children. This is bound to influence their parents' perspective on the matter. Second, although I have spoken of communities that 'have lost their school', in fact no area ever ends up without a school altogether. The children are always sent to an alternative school that

in time will also be seen as 'our school', whether or not
relations with the new school are ever as close as they
were with the old one. And third, although the use of death
imagery is very frequent with reference to school closure
and effectively expresses people's fears about the process,
the fact is that communities do not die in the way that
individual animals do: to speak as though they do is reific-
ation. Except in the case of fatal catastrophe or complete
outmigration, communities simply change through time. Such
change is normal, although not always welcome. Although we
approached the study areas with every sympathy for the views
of small school supporters, it would not be accurate to
report that any of the communities had died as a result of
school closure. After the fact, not a single respondent
suggested that this had taken place, nor did we hear of
families that had moved away because of a closure. As
far as we could tell, although local people were not happy
to have their schools closed, in any sense in which it is
meaningful to apply the notion of life or death to a
community, these communities survived.

But if the data do not support the notion that school
closure is 'fatal' for rural communities, there is also
little evidence that closure brings about any social benefit.
The most positive conclusion I can offer on the basis of
this study is that rural areas seem to survive closure,
but at a certain cost both to the children and to members
of the community in general. As this study has shown, this
cost is considerably more complex than the loss of a building
to meet in or the departure of a teacher with presumed
leadership potential. Rather, closure removes from the
community a means of social integration, a reason for and a
channel of communication, and a symbol of community identity.
In some cases, some other institution may take over one or
more of these functions, or indeed may already be performing
them. In others - particularly where other participatory
institutions such as the church have already been closed -
these functions may cease to be performed. This is without
doubt a loss to the area.

Moreover, school closure against the wishes of local
people seems to increase their feelings of powerlessness
and alienation with respect to the organs of local government, and can lead to enduring resentment. These changes
do not add up to the destruction of a rural settlement, but
I suggest that they do constitute social damage. Certainly
they are serious enough for us to conclude that for social
reasons alone, policy-makers should never make a school
closure decision lightly.

I began this paper by referring to certain phrases in
common use which express the idea that schools are the
central institution in rural communities, and that closing
the school condemns such a community. I then described some
research showing that although people in rural areas of
Scotland do not welcome school closure, the social effects
they perceive after the fact are much less dramatic than
those people anticipate.

If school closure does not in fact 'kill' rural communities, then why do people resist closure so strongly? In part, no doubt, they resist because they believe the commonplaces. But I think that another factor is involved here as well - the feeling of powerlessness mentioned earlier. People seem to perceive a closure proposal as an attack on their community, as something that 'they' (outside officials responsible for the formulation and implementation of public policy) are doing to 'us'. Resistance to such proposals expresses not only ideas about children and schools, and the perceived right for a settlement to continue as a 'community' distinct from neighbouring settlements; it also expresses people's anger at having important decisions made about their lives without their consent. Of course, such decisions occur quite frequently, and affect other areas of life besides primary education. But I suggest that for many people, school closure decisions have come to symbolize the erosion of people's control over their daily lives. Resistance to school closure thus expresses resentment of and resistance to that wider process as well as to the prospect of closing any particular school.

If this explanation is valid, then resistance to school closure might be seen as an expression of localism - an increase in concern with local identity and autonomy in the face of growing encroachment by the state. While this may explain much of the motive behind resistance to school closure in rural areas, it is ironic that the language people use to express that resistance is so highly stereotyped. When a rural school is 'defended' with such phrases as 'the school is the heart of the community', that school's importance is being established not on the basis of specific information about its contribution to a particular local culture and social setting, but rather in terms of general beliefs about the local-level importance of rural schools. Such argument may be localist in intent, but it is hardly local in content. It rests on terms external to any particular local reality, on national rather than local culture.

NOTES

[1] The House of Lords debated "Village Schools: Government Policy" on 19 March 1980. See *Parliamentary Debates (Hansard), House of Lords Official Report,* vol. VO7, no. 101, cols, 307-339. See also the letter page and editorial columns of *The Times* for September 1978, including the letter by Lady Plowden on 7 September and the editorial on 11 September.

[2] The research described here was carried out at the University of Aberdeen, where I was a Research Fellow in the Institute for the Study of Sparsely Populated Areas and the Department of Education from 1979-81, and in the Institute for the Study of Sparsely Populated Areas from 1981-82. The work formed part of a research project funded by the Scottish Education Department, the Scottish Development Department, and the Department of the Environment; it is reported in Forsythe *et al.* (1983). I was assisted in the interviewing by Rosemary Lumb, Eleanor McNab, Cynthia Sadler, Peter Sadler, John Sewel, Derek Shanks, and Jennifer Welsh. The views expressed in this paper are those of the author and not necessarily those of other members of the research team or the funding agencies.

[3] OECD's Centre for Educational Research and Innovation recently (1979-81) sponsored a major international project on Education and Local Development in sparsely-populated areas. This project was designed to "look at the roles (actual and potential) education plays in the overall development of the local communities being served". (Sher 1981, p. xvi).

[4] In this report, 'non-parents' refers to respondents who had no children in primary school at the time of interview. Some of these people did have children above or below primary school age.

REFERENCES

Forsythe, D., 1983a. A review of previous research, in:
D. Forsythe *et al. The Rural Community and the
Small School,* (Aberdeen: Aberdeen University
Press) pp. 15-49.

Forsythe, D., 1983b. A comparative survey, in D. Forsythe
et al. The Rural Community and the Small School,
(Aberdeen: Aberdeen University Press), pp. 83-136.

Forsythe, D., and Carter, I., 1983. The local view: a case
study approach, in: D. Forsythe *et al. The Rural
Community and the Small School,* (Aberdeen:
Aberdeen University Press), pp. 137-170.

Forsythe, D., Carter, I., Mackay, G.A., Nisbet, J., Sadler,P.,
Sewel, J., Shanks, D., and Welsh, J., 1983.
The Rural Community and the Small School,
(Aberdeen: Aberdeen University Press).

Jones, H. (ed.), 1982. *Recent Migration in Northern Scotland,
Pattern, Process, Impact,* (London: Social Science
Research Council, North Sea Oil Panel Occasional
Paper no. 13.).

Lauglo, J., 1982. Rural primary school teachers as potential
community leaders? Contrasting historical cases
in western countries, *Comparative Education,* 18,
pp. 235-255.

Lumb, R., 1980a. *Migration in the Highlands and Islands of
Scotland,* (Institute for the Study of Sparsely
Populated Areas, University of Aberdeen, Research
Report no. 3).

Lumb, R., 1980b. Communicating with bureaucracy: The effects
of perception on public participation in planning,
in: R.D. Grillo (ed.) *Nation and State in Europe,*
(London: Academic Press), pp. 105-117.

Murray, J. and MacLeod, F., 1981. Sea change in the Western
Isles of Scotland: the rise of locally relevant
bilingual education, in: J.P. Sher (ed.) *Rural
Education in Urbanized Nations: Issues and Innovations,*
(Boulder, Colorado: Westview Press), pp. 235-254.

Sher, J.P. (ed.), 1981. *Rural Education in Urbanized Nations:
Issues and Innovations,* (Boulder, Colorado: Westview
Press).

Solstad, K.J., 1981. Locally relevant curricula in Rural
Norway: The Lofoten Islands example, in: J.P. Sher
(ed.) *Rural Education in Urbanized Nations: Issues
and Innovations,* (Boulder, Colorado: Westview Press),
pp. 301-324.

13. Images of place in a Northumbrian dale

BRENDAN QUAYLE

INTRODUCTION: THE SETTING OF 'PLACE'

On commencing fieldwork in the Allendales in 1981 one of
the strongest early impressions I gathered was the apparent
obsession of the local people with 'place', with the state
of their immediate physical environment. So as a preliminary exploration of Allendale local culture I examine here
certain aspects of this interest in, and concern for,
place. From the ethnography set out it will be seen that
these attitudes are underwritten by a strong local 'sense
of place', a set of culturally established values and
meanings which arise from the experience of belonging to or
'being a part of the place'. Indeed, I would argue that in
the Allendales 'place' provides both the symbol and framework of community. Attachment to place not only underlies
and transcends the local sense of community, it also
establishes the basis of community itself.

    To illustrate the importance of place as a key construct
and unifying idiom within Allendale local culture, I examine
some local responses to certain changes which have been
proposed for the area, and to others which have already
taken place. In these responses can be seen the workings
of place in local consciousness as an idiom of self-
reflection, as a source of symbols and images for expressing
ideas about both the contemporary nature of community and
its current state of health.

    Local identification with place, both on the part of
individuals and entire communities, can be seen as an
aspect of symbolic classification which is directly comparable
to the homologous relationship between community and its
physical environment that is widely represented in traditional
cultures of the type more usually studied by anthropologists.
In tribal and Third World societies the physical world,
place, object and space are commonly deployed as idioms for
the conveyance of ideas and cultural norms. They provide
a vocabulary, a lexicon, for the transmission of social
structure: they reflect that structure and reinforce it

simultaneously. In totemic societies, such as those of the American Indians or the Australian Aborigines, the natural environment itself provides the primary medium for communications of this type: it serves as both model and mentor for social organisation and patterns of behaviour (cf. Levi-Strauss 1969; 1972a). And in these and other societies, artificial space, the layout or positioning of a settlement, or buildings within it, is invested with the key organisational and ideational principles of the culture in question (cf. Levi-Strauss 1972b; Cunningham 1973; Bourdieu 1971; Hocart 1970, pp. 250-261).

In this instance, we are not dealing with such 'total' systems of classification, involving multiple correspondences between physical, social and metaphysical domains, but the ethnography I set out here does demonstrate the persistent symbolic value of place within the traditional and contemporary rural milieu. Place within the local culture of the Allendales may not mirror local social forms; but it does function as a focal point of attention, identification and discourse; it is used as a primary medium for expressing and re-inforcing key community constructs and rural values, even within a context of change.

The Allendales are an upland farming region of south-west Northumberland. They skirt the borders of County Durham to the south and the east, and Cumbria to the west; and they consist of two valleys, East and West Allendale, formed by the two branches of the river Allen, a tributary of the South Tyne. Once united as the "Ancient Parish" of Allendale, they now form three civil parishes, Allendale (population 1445), West Allen (276) and Whitfield with Plenmellor (197). Allendale Town and the village of Catton, lying within a few miles of the junction of the two rivers, provide the main centre of population in the area (1086). The remaining population inhabit farmsteads scattered around the dales, but there are small concentrations within the six villages and hamlets which lie at intervals along the valley floors.

The villages of Allenheads (160) and Whitfield (137) are estate settlements, the focal points of two large estate organisations which together own over 16 000 ha of land in the area, over half the land in the dales. The rest belongs to private family farms, the majority of which were bought originally off the estates. The wealth of the estates was founded upon the lead mining industry which dominated this and adjacent parts of the Northern Pennines until the end of the last century. But the economy is now wholly agricultural, a mixture of sheep farming, dairying and beef breeding, managed both by tenant farmers on the estates and free-holders. The income which the estates derive directly from rent and agriculture is supplemented by the leasing of grouse moor to shooting consortiums, and by the felling and planting of timber. The area does attract regular tourists, but the numbers are not sufficient to make much impact on the local economy: two of the largest hotels in Allendale Town are only open for half the year.

SINDERHOPE: THE DEVELOPMENT OF PLACE

Like most rural communities, the settlements of the Allendales have each a conspicuous centre of place, a symbolic
focal point such as a green, or market place, near which is
usually located the focal institutions, the school, church,
pub and post office. And this is true of both lineal and
nucleated villages, though the 'centre' may be actually
located to one side of the village; on a roadside, or at
the crossroads.

One place in the area, however, because of its physical
configuration, lacks any kind of obvious centre, any marked
or set apart focal point. This is Sinderhope, a village
with a substantial population of 85, but consisting largely
of fourteen square miles of fell and fields dotted with
farmsteads spaced at varying intervals from one another.
Strung across the roadside however at half mile distances
are the three buildings which identify the area as Sinderhope, a disused Methodist chapel, a tiny sub-post office
and a schoolhouse. Between them lie entire fields and
ditches, a rivulet and a roadbridge, and a couple of roadside dwellings, one a former joiner's shop and the other a
former police house.

Over the years, the ease of access to nearby Allendale
Town and the diffuse nature of habitation in Sinderhope
has led to the locals looking outside of the place for their
social activities, and for their point of orientation
within the dale. However, in 1979, a dispute about 'place'
occurred in Sinderhope which effectively revived the local
sense of community and brought about a re-orientation of
place back towards Sinderhope itself. The goal of the
local party to the dispute, and the culmination of their
endeavours, was the restoration of Sinderhope schoolhouse
as a village institution, and its investment with a new
symbolic function as the focal 'centre' of place. In the
sections following, I outline the salient facts of this
dispute, and I detail the range of meanings, both explicit
and implicit, which underlie much of what took place.

THE SCHOOLHOUSE DISPUTE

Sinderhope school was closed down in the 1950s by the local
education authority. The school building was converted
into an 'Expedition Centre' to provide outdoor activities
for urban school-children. The local children were sent
out to Allendale Town and to Haydon Bridge to be educated.
Closure of the school was much resented in the area. Local
people thought it ironic and unfair that their own kids,
who they felt "belonged to the place", had to be sent out
of the place to the towns to be schooled, while kids from
the towns were brought into the village to be educated in
a rural environment.

However, in 1979, the County Council cut expenditure on these expedition centres; and they decided to close Sinderhope and a number of the other converted schools in the area. The village learned that the council were to sell the school and divert the funds elsewhere. This provoked immediate opposition from Sinderhope residents who knew that the school had been built in 1856 out of local public subscriptions raised and supplemented by the Beaumont Lead Company. Through parish councillors and letters to the local newspaper, the *Hexham Courant,* the morality and the legality of such a sale were disputed: they had no "right", it was claimed, to dispose of a community asset, like the school, for "gain". Raised by local money, it essentially "belonged to the local community" (Letter, *Hexham Courant,* 6 April 1979).

In response to this community feeling, and as a well-meant gesture to stop the school ending up as "just another weekend cottage" - another much resented feature of local decline - a prominent local councillor and dale 'incomer' proposed that it should be converted into a recreation centre for skiers and locals, and that land adjacent to the building be used to house an all-weather, plastic, dry ski-slope. A representative from the English Ski Council surveyed the site and pronounced it ideal for a skiing centre. But when this scheme was aired at a public meeting, the local people, rather than viewing it as something which would bring money into their area, and thus provide jobs and capital to maintain the school as a public facility, saw it as an "idiotic scheme", a potentially "disastrous" plan.

The subsequent outcry was more intense than the original opposition to the sale of the school. Residents argued that the ski slope would be "unsightly"; that it would attract a volume of unwelcome "strangers" to the area; and that it would bring only "nuisance and vandalism". The use of the building by skiers, they also felt, would leave no room for local activities; again control of their school building would be taken from them. These complaints were well publicised in letters and interviews in the local newspaper; and it was not long before the north eastern regional dailies, the *Evening Chronicle* and the *Newcastle Journal,* sniffing a feud between local residents and county bureaucrats, gave it headline coverage. One story ran "Wrangle over the old village school: BID TO STOP A SKI RUN IN ITS TRACKS" (*Newcastle Journal,* 8 May 1979).

Feelings were running so high by this point that the residents of Sinderhope set up an action committee to "make formal public disapproval of the proposed ski centre" and to draw up an action plan to save the "Centre" for the local community. The committee comprised nine local residents - five farmers, the local policeman, the post-mistress, a commuter and a lead miner. Two were 'incomers', the rest were from families of long local standing. They met seven times between April and June of 1979, to organise

a petition opposing the ski centre and to discuss the
results of a feasibility study into the school's potential
as a museum and as a community centre. They sought planning
and legal advice; co-opted practical assistance from the
Northumberland Community Council and from British Steel;
and joined forces with the High Forest Association, a
community organisation based in the upper dale which ran
local concerts and the local flower show.

The survey of local attitudes to the idea of a ski
centre produced a 95 per cent local response against the
plan. Armed with this level of community support, the
committee lobbied local councillors and local candidates
for the forthcoming general election, eventually winning
over the Parish Council and the District Council, but not
the local MP (who had a reputation as a fence-sitter) to
their side. By September, after a barrage of letter
writing, petitioning, complaints to the local press and
lobbying by parish and district councillors, the County
Council gave way on most counts. Not only was the idea
of the ski centre dropped, but also the proposed sale of
the school building. The committee were offered the use
of the building at a nominal rent of £10 per week on a
two year lease with possibilities of extensions, depending
upon the extent of its use and the condition in which it
was kept.

At a celebration meeting, the school committee agreed
the terms offered by the County Council, and handed over
the running of the building, now renamed the 'Sinderhope
Community Centre', to the High Forest Association. In a
short space of time it was converted with local funding
and voluntary labour for community activities. A bonfire
party for the Sinderhope children in November was the
inaugural event. And this was followed by an extensive
weekly programme of social functions, indoor sports, playgroup activities, talks and shows. The school was once
again in the hands of the local community.

ANALYSIS: A QUESTION OF BELONGING

To understand why the people of Sinderhope were prepared
to put so much effort into retaining the school for local
usage and prevent it falling yet again into the hands of
'outsiders', two interrelated factors must be considered:
firstly, the total inappropriateness, in terms of local
cultural values, of the uses to which the school would be
put, once its future was taken out of their hands; and
secondly, the local evaluation of institutions like village
schools as vital 'centres' of place, not just focal points
in dale topography, but as symbols of a communal and
domestic continuity of culture which straddles change and
the passing of generations. Nobody believed that, if
their campaign proved successful, it would be restored as
a school; but they would at least have secured control of

a focal point of place which 'belonged' to the community; and the idea of using it as a community centre would go some of the way towards fulfilling its former practical and cultural functions. It would re-create the social inter-change within the home area necessary to preserve local identity; and it would, hopefully, encourage the young to socialise within, and identify with, that home area.

To an extent however, interest in the actual usage of the school as a centre for local activities and social life was incidental for many of the campaigners, including over half the action committee itself. The practical need for such a building was never really an issue. Facilities for shows, plays, social evenings and sports already existed at the Beaumont Hall in Allenheads, the traditional meeting place of the upper dale Women's Institute and the High Forest Association, and at the parish hall in Allendale Town. Neither place was more than three miles from Sinderhope. At issue was a question of principle, the disputed right of an external, and to some a remote, agency the County Council, to dispose of one of their historical institutions above their heads. And overlying the matter of identification was the fact of ownership; they believed that morally and 'by right' it belonged to them.

The initial fear was that the building would become a holiday cottage, adding to the already growing list of local farmhouses and estate cottages owned by strangers and occupied only at the height of the summer months. The holiday cottage 'problem' - rows of ancestral homes lying empty and lifeless for most of the year - is believed to have turned the villages of the upper dales, Allenheads and Carrshields, into 'dying villages'. And there are fears that this form of social death will spread down into the lower dales. The estate at Whitfield in the lower West Allen refuses to let holiday cottages on any account; and in Allendale Town opposition to such lettings is very strong, both among residents and local councillors. One local resident saw the planned sale of the school as a manifest-ation of council policy, which in his eyes, was to quicken rather than impede the process of rural decline: he said of the decision, "they want tae drive us oot, then they wonder what th' dreft is fram th' rural areas".

The issue intensified considerably when the suggestion of the skiing centre was raised. The underlying motive was practical and prudent: the skiers would bring in the money necessary to develop the building as a community centre; and thus prevent it becoming a holiday cottage. But the proponents of the scheme, backed by the recreation committee of the Tynedale District Council, had failed to reckon with the strength of local feeling against this form of tourist activity. In the upper dale friction had been building up since the early 1970s between the villagers of Allenheads and the large parties of suburban skiers who come from Tyneside every year to utilise a ski run leased out by the Allendale Estate to the Norwegian Ski Club. The run was sited alongside an approach road into Allenheads, on an area of fell lying above Allenheads Hall and the centre of the village. To the villagers, the usual problems which beset this high-lying community during the annual snowfall, have been regularly aggravated by the additional problems brought by the skiers. They "swamp the place"; and the centre of the village is turned into a sea of cars, blockages and churned up mud and snow.

Often appearing insensitive to local needs of access and parking, and seemingly oblivious to the existence of a community in Allenheads reliant on certain essential services, the skiers are regarded by the villagers as a complete "nuisance". "You can't get in or out", said one villager. "They're no good to anyone", said another, "they bring no good to the place". Often showing no concern for local privacy, they regularly rile the locals. One year, 1971, a row broke out which made headlines in the regional press: "SKIERS BRING LIFE TO A STANDSTILL" (*Newcastle Journal*, 8 January 1971). A villager, prevented from making a mercy call to a sick relative in another dale, was quoted as saying "They just stood and watched me trying to get through and no one would move".

At one level, the annual conflict between the villagers and the skiers, is a conflict between rural and urban attitudes. For the skiers, Allenheads is a playground stocked, it would seem, with antediluvian bumpkins, who in the jibes of one visitor, hadn't "got beyond the stage of the horse and cart". But for the locals, it is their place, their community, their amenity and their source of livelihood. In their eyes, the skiers only bring "chaos" to the sacred centre of place. The sea of cars and ski suits pose a threat to the character of the village; its traditional appearance and functions. The custom of the skiers, beneficial to no-one except perhaps the publican, is unacceptable to the locals. Their lack of interest in the locality itself, and their lack of deference to local values and needs makes their business inappropriate to the area: ill fitting to local character of place.

The villagers of Sinderhope were anxious that this legacy of urban/rural conflict might spread down the valley into their place. Had the idea of the ski slope, an "idiotic" scheme in the words of one observer, been imposed, it would have brought unacceptable levels of practical and visual disruption into Sinderhope, not just for one or two weeks, as in Allenheads, but all the year round. Not only would the new wave of skiers represent, in the famous idiom of Mary Douglas (1970), "matter out of place"; but the accompanying developments would have been, in their terms, inappropriate to the place, detrimental to its rural character, and a danger to the identity of the local society attached to that place. Hence tha almost unanimous opinion, widely quoted in the local press, "we don't want a ski slope here".

THE 'CENTRE OF PLACE' AS A SYMBOL OF COMMUNITY

Some of the images of desecration and intrusion of place conjured up by the villagers of Allenheads in describing the activities and attitudes of the detested skiers are founded upon certain, specifically local, cultural evaluations of the centre of place. The school at Sinderhope was one such centre of place. Its closure, like that of all the village schools in the valleys, was seen locally as the prime cause of local decline: "The first nail in the rural coffin" as one commentator put it. "The worst thing that ever happened to the place" said another, lamenting the absence of children's voices and the resultant loss of dialect and commitment to 'the place'.

As a rule, the best tended part of all the dale villages and hamlets is always the very centre of the settlement, its green or market place and the institutions which surround it. This pride of place goes deeper than simply winning prizes in the Best Kept Village competition. For in local terms, the external condition and appearance of the village centre is seen as a reflection of the internal state of the community which inhabits it. The most desirable place to live in the whole area is Catton, the village with the most orderly centre in the dale. This is a community of the well-to-do, of commuters and retired farmers, whose material well-being and well-organised community life is echoed in its tidy, well-manicured greens and churchyards, and in the neatly trimmed hedgerows and gardens of its most prestigious residential area, the so-called 'golden mile'. The least desirable is the remote village of Carrshields, a once thriving mining and agricultural community, but now rapidly depopulating and said to be 'dying'. The remaining villagers complain of Carrshields as having 'lost its centre'. The school and post office are closed; nearly half the houses are holiday cottages and the church and church hall where social activities were once carried out have been demolished.

Midway between the extremes of the Catton and Carrshields centres, are the village squares of Allendale Town and Allenheads. Parts of these are said to be 'badly kept', a complaint aired regularly at Parish Council meetings, and the subject of a number of local disputes and conflicting interests. Because the centre of Allenheads is owned by the Allendale Estate, which refuses to maintain it on the grounds of the expense involved, neither local residents nor the parish councillors feel able to do anything about it, other than submit annual complaints to the estate agent. The estate's lack of care is bitterly resented: residents view it as a symptom of the estate's lack of interest in the welfare of the local community whose work in former days made possible the wealth of the estate.

The problem in Allendale Town is also one of conflicting interests, though in this case these are wholly internal. A busy local garage uses large parts of the square as a parking and vehicle display area. Local residents, particularly those whose homes adjoin the square, consider the rows of land rovers and parked lorries as an intrusion, as detrimental to the 'spirit' of the place, its aesthetic appeal and its peaceful rural character.

In contrast to the acrimony arising out of its everyday usage and appearance, is the utilisation of Allendale Town square every New Year's Eve as the site of the 'Tar Barls' procession and bonfire. This traditional dale festival, performed by locals for and on behalf of the local community and its visiting dale emigres, occupies a focal position in the local culture of belonging. The route of the procession of fiery tar barrels circumscribes the town centre, the turning points provided by the gateways of the village institutions, the two churches, the two schools and the parish hall. Just as it thus maps out the symbolic spatial heart of place, the festival also acts as a registration and re-affirmation of the authentic local value of community and its association with the place; for participation is a hallmark of local identity. As I have shown elsewhere (1981), the event provides the town and the surrounding dale with its most powerful image of centre, dramatically linking the physical heart of the place with the cultural heritage and sense of belonging which lie at the nucleus of the local ancestral community.

Taking into account the emphasis in local culture given to the centre of place as a symbol of community, the fight for Sinderhope school must be seen as a struggle by a community threatened by decline and diffusion to re-define itself anew through the restoration of its physical and symbolic centre, in this case the old village school. The demise of the chapel and the former expropriation of the non-denominational school for outside interests represented a loss of centre, and meant an inevitable, and locally detrimental, diversion of interest and attachment away from place, to areas outside Sinderhope. In coming together to prevent the complete loss of the school, both insiders and outsiders living in the place acted out a local cultural prescription, the use of place as an idiom for forging and conserving the value of community.

'PLACE' AND CHANGE

The outcome of the Sinderhope dispute was also a victory for the shared conservatism of both local and incomer, a repudiation of local development and change in keeping with previous opposition to similar schemes. To date, these have included: a plan to turn the area into a National Park (1971); a proposal to designate the North Pennines, including Allendale, as an Area of Outstanding Natural Beauty (AONB), to promote conservation and informal recreation (1972); an assisted tourist growth scheme, part of a North Pennines Growth Point programme, underwritten by the

Tourist Board (1978-81); and a variety of schemes to
encourage farmers to lease fields for caravan sites and
provide farmhouse accommodation for visitors.

Ironically, schemes which few conservationists or
recreationalists would approve of, such as quarrying, mineral
extraction, goods manufacturing and large scale vehicle
haulage, find favour in the area, and suffer little resistance from councillors and planners. Underlying this is the
neo-industrial legacy of Allendale as a former lead mining
and processing area. Large parts of the local landscape,
and the shape of the dale villages, owe their forms to lead
mining processes. The legacy is still ingrained in local
cultural forms, from place names to kin links, from dialect
to the pattern of agricultural smallholding.

Equally, the area's rugged beauty and its highland air
have always attracted visitors, and tourist provisions have
existed since the middle of the last century. But in
cultural and economic terms, tourism has always come a poor
second to mining and quarrying, with the close interdepend-
ence between the latter and local agricultural concerns and
practices. Until recently, the area supported four such
labour-intensive, mining and quarrying operations and one
advance factory - without any opposition from the local
communities.

Again these attitudes are underpinned by the local
perception of what is and is not culturally, as well as
economically, appropriate to the area, to the place as a
whole. However, in recent years, the Allendonians have
witnessed the demise of all local mining enterprises, and
of the estate which financed its agricultural operations
from these. At the same time, they have observed the
inevitable encroachment of the new recreational economy,
in the form of hobby farming, expedition centres, holiday
cottages, outward-bound and pony trekking establishments,
ramblers and guided parish walks; and the advent of various
groups retreating from perceived urban madness to the
apparent sanctuary of the wild hills, including the
vegetarian potter, the artist-recluse and two oriental
religious retreats.

However, the changes which have accompanied the march
of this new economy, occurring in parallel with the last
throes of the old economy, have given rise to a new sense
of place, of their home area as a place of decline, of
social death and decay. The new economy in particular,
with its cosmopolitanism and its lack of deference to local
culture and identity, is largely viewed as an aspect of
that decline, of hastening the eclipse of place as it is
known and valued. This process is a familiar one in rural
parts though perhaps the imageing of tourism as decline is
possibly specific to the Allendales.

This considered, I want to turn now to an examination
of the way that the local consciousness of decline has itself

become reflected in the changing perceptions of the Allendonians of their wider physical surroundings, both natural and man-made. For the ethnography I move away from Sinderhope and Allendale Town to Allenheads (population 160) at the top of East Allen Dale.

ALLENHEADS: THE DECLINE OF PLACE

When the people of Allenheads volunteer to the ethnographer the information that their village is 'dying', they are making a statement about its social death: the lack of employment; the absence of children; the domination of the population by the elderly and the retired; and the decline of social intercourse which afflicts an ageing, less mobile community. But apart from the inevitable comparison with the 'old days', when there was believed to have been more 'life in the place', such things are rarely talked about explicitly in general conversation. Instead, this sense of inner decline, is projected onto their immediate environment, the village and the countryside around. The perceived physical decline of the place provides a powerful image through which they articulate feelings and experiences of economic and social decline.

Older villagers point to the ruins that lie within the village, just off the village square, the smithy, the old granary, and the old storehouses of the miners; and they complain of the neglect which has allowed their walls to crumble and the undergrowth to take over. This chaos in the heart of place is described as a 'return to wilderness', as a sign that the village is 'going to ruin'. In the same breath, they describe the margins of the village as 'going back to the fell', going back to the chaos of the wild, inhospitable hills. Evidence of this is provided by the presence of fell grasses and rushes inside the village walls, the decaying gardens of the elderly and the recluse, crumbling walls and broken fences, and an epidemic of moles and rabbits that have encroached within their domestic spaces.

The appearance of this chaos of wilderness within the heart of place is ascribed to a positive policy of neglect and lack of care for the village on the part of the estate which owns the houses and the land around. The landlord's indifference to the place is equivalent to his indifference to the people who live in and identify with that place. Just as it is said that the place has been allowed to 'go back to the fell', the older tenants also cynically joke that 'now we're old and retired, we've had it now, finished up, they're just waiting for us to go'. The estate, it is felt, views a vacant cottage as either a diminished burden or, if it can be sold, as a source of ready capital which can be diverted elsewhere, out of the estate. The sale to outsiders of local homes which in some cases have housed local families since the eighteenth century, is perceived, like a death, as a loss to the community. With each new sale, villagers openly express their fear that, of them and their families, there will soon 'be nothing left'.

The well-being of Allenheads, both the people and the place, has always been closely tied up with the fortunes of the estate and the traditional land-owning family, the Blackett-Beaumonts, Lords of the Manor of Hexham, and now based at Bywell in Hexhamshire. The estate is run ('run down' is the quip) by an impersonal agent who operates at a distance; and Allenheads Hall is let out as a shooting residence to wealth American consortiums. It once had 18 hall staff and, until the mid-1960s, another 18 ground staff. But now only the five gamekeepers who care for the grouse-rich fell and its remunerative shooting rights are kept on full-time. Minimal maintenance of the hall, its grounds and, until recent cuts, the village buildings, is carried out by a handful of retired estate staff, now working part-time.

Traditionally, the villagers identified totally with the estate, as their livelihood and their place of residence; they 'belonged' to it. Equally, the estate was wholly assimilated with the landlord whose name it bore. A duty to the estate was a duty to the landlord and vice versa; a correlation of place and its leading figure which is still found in neighbouring Whitfield, where in local parlance, the 'Squire's heart' still 'beats in the place'. But now that the present Lord Allendale has wholly removed himself from the estate, its symbolic status as a focus of belonging has diminished considerably. The gradual with-drawal of his presence over the years corresponds in the folk view, to the physical decay of the estate. His now total absence from the place predicates the expected imminent extinction of the estate itself. As one villager casually joked, "when I woke up this morning, I looked out to see if it (the estate) was still there". Perhaps, inevitably, comparisons are made with Whitfield, the estate in the West Allen, where the resident landlord 'keeps everything in place', i.e. it is orderly in appearance, well managed, fully staffed and prosperous. One resident who had lived and worked for most of her life first in Whitfield and then in Allenheads described the contrast:

> (Whitfield) was a nicely kept place,
> everything in its place, such a pleasant
> village ... with the squire on the spot
> and there to see everything was kept just
> so. But ye never saw any tumble-down
> walls like there is here or anything like
> that. I've been here twenty three years
> and I've never seen any place deteriorate
> in my life like this has in those twenty
> three years. When we came there was a
> host of estate workers, there wasn't a
> wall down or a stone out of place, and
> there was gates on those yards, and

                now ... look at it; and no workers at
                all. (Allenheads) is rapidly becoming a
                village of pensioners ... it'll be a
                dying village. And from here to the
                Holms (near Allendale, the perimeter of
                the estate), my husband used to cut all
                those hedges ... I don't think they've
                ivver been cut since.

                I can't tell you whether its lack of
                money, or if its just lack of thought
                for the place. Because this estate up
                here now, its the second son's ... and he
                does nothing.

Formerly, the estate defined the nature of community,
as it still does in Whitfield (cf. Quayle 1982) and the
landlord and his place of residence, the hall provided the
source and the symbolic epi-centre of community itself. In
his absence, the hall has become only a 'shooting box',
a kind of manorial holiday cottage. With their diminished
status, their perceived 'loss' to the place, the local
sense of community has itself become fragmented.

This receives further confirmation for the locals in
the gradual physical fragmentation of the estate itself,
the disposal 'for gain' of the very asset which has traditionally underwritten their livelihood and their experience
of belonging. This, and the frequent sales of local farms
to outsiders and the hobby farmers and of village cottages
to holiday makers, are all highly emotive issues in Allenheads. For the older residents, it is as if the ground
under their feet where their families have stood and worked
for generations, has been taken away. This fragmentation
inculcates a further sense of community insecurity; it is
interpreted as leading inevitably to the physical dispersal
of family and community. Lacking any real knowledge or
insight into estate plans and policies, the future is
uncertain.

But perhaps an even more pungent image of death and
decline in Allenheads is provided by the fate of the
Beaumont Mine in the village. Prior to its initial closure
in 1896, this mine had yielded over 300 000 tons of lead
and silver ore for the Blackett-Beaumont Lead Company, and
it effectively created the wealth of the estate. It lay
dormant until 1972, when British Steel leased the site from
the estate and started mining for fluorspar. Over two
million pounds was invested; and a new stone-built mineyard
was constructed over the old workings. Water was pumped
from old shafts which had lain flooded for over half a
century. Some 22 local men were taken on to operate the
mine. Optimism was high. The village institutions, pub
and post office looked forward to the new business the
mine would bring. The regional newspapers described it as
a "gold mine". And despite scepticism on the part of
older lead miners in the area as to the true extent of the

mineral reserves, the village as a whole looked forward to a new lease of life. But the old miners were right, and in 1979, after it bacame clear that the predicted deposits were lacking, the mine was closed.

Its closure was felt as a cruel blow within the village. Its promise of regeneration had come to nothing. Villagers talk about the 'waste' and the 'shame' of the mineyard standing there, idle and empty. References are made to the 'mess' created by the iron girders lying in rusting heaps and also to the invading bracken and mosses. The presence of decay and the atmosphere of total stillness provide for them a powerful image of the acceleration of village decline and the inevitability of village death. As one resident remarked, "All day it used to be a clang and a clatter in that yard the whole time, you know; and stuff was coming up out of the mine. And now it's all silent except for the pumps ... It's been awful since they closed it down, for silence".

Formerly, the mine provided work and a sense of prosperity, now there is just an 'emptiness'. This too is a recurrent image which constantly surfaces in conversation and dialogue. And the 'emptiness' above ground - in the absence of people and of activity - merges in the imagery of place with the emptiness which lies below: a few days after the pumps were stopped (in 1981), confirming the unlikelihood of any revival, a gaping hole appeared in a village back garden - a shaft below had caved in. Comments were passed around about the village being "riddled full of holes" because of the extensive tunnelling. After all, it was said, who would really bother if it, or parts of it, were to be engulfed in one major subsidence. As there was "nothing in the place now", who would "notice"?

The images of physical decay, of ghostly silence and of emptiness, all contribute towards an overall definitive image of the village as a 'dying' community; a state of social and economic decline perceived and expressed through a vivid imagery detailing the death of 'place', the community's physical environment. Allenheads has "all gone dead now ... and it'll just go deader, until there's nothing left ... it's a fact".

PLACE AND BELONGING

In this paper I have been interested primarily in the way that place is used in Allendale as a primary vehicle for thinking about the changing nature of community and the changing qualities of community life. Some of the compara- tive literature on the culture of belonging touches recurrently on the imagery of place and the notion of 'sense of place' as expressed in local culture. The volume edited by Anthony Cohen (1982) is representative of work in this field. Cohen's own analysis of close social association

on Whalsay uses 'sense of place' as a way of describing
the islanders' identification with their place of primary
kin residence, a place which ultimately merges with the
island itself as the physical location of an entire interlinked community. Place, he shows, is used in Whalsay
folk culture, together with time as a 'vocabulary' for
expressing local attachments and associations. The Whalsay
folk model, it appears, makes no clear cut distinction
between the community and its physical setting. Whalsay,
the island, provides the community with both its physical
and its social parameters. Equally, the various parts of
the island are used to identify not simply the place of
origin of the various community members but also the
territorial bases of particular units of kin.

This correspondence between community and place also
crops up in a brief description by Emmet in the same
volume on 'sense of place' in the Welsh town of Blaenau
Ffestiniog (1982, pp. 203-205). In this she shows that
despite various obvious disadvantages, it is still a place
in which people who could choose to leave, prefer to live.
Certain perceived advantages, 'enjoyment' and sentiment
only account for part of this preference. Attachment to
the town, its physical character and its environmental
setting is synonomous in the consciousness of the individual
townspeople with their attachment to the local community,
its language and history.

Another contributor, Mewett (1982, pp. 222-246), in
a discussion of the social and symbolic significance of
migrants and migration for the inhabitants of a Lewis
crofting community, shows the part played by sentiment in
promoting return migration and thus providing "a relational
base for a local consciousness" (p. 240). A critical part
of this sentiment is 'attachment to place', meaning by this
"that people retain an emotive attachment to their homebased
social relationships which is translated into an attachment
to place" (*ibid*), to belonging to a named village like
Clachan.

Finally, in the same collection of papers, Strathern
introduces the notion of place, in this case the village
of Elmdon, as fundamentally an idea, an idiom distinct from
the status of the village as a natural entity. She differs
from the other contributors in seeing the equation of
place (village) and community as being unwarranted.
Elmdoners' notions of belonging to place, she suggests,
must be understood in terms of a wider field of ideas within
English society, including class values and received notions
of village-ness (p. 249). In any case, within Elmdon itself,
the influx of 'outsiders' has led to differential perceptions
of 'belonging-ness' within the village, and to the elaboration
of distinct categories of inhabitants to whom are assigned
differing degrees of attachment to place in alignment with
their kinship affiliations.

However, in Allendale, it is no longer simply kinship
and birth, or even extra-local, class-based notions of

village-ness which establish the boundaries of local community membership. Rather, this has to do with association with place (over a respectable period of time) combined with an exhibited degree of commitment or deference to that place. Asked directly about the terms of community membership, informants maintained that the community includes both those who 'belong' there by virtue of birthright, dialect and membership of one of the old Allendale families, and those who come in from the 'outside' and play their role, through work or voluntary effort, in maintaining the 'life of the place'. As a mode for regulating the nature of local community and defining the terms of membership, attitude to place has priority over kin ties and authentic 'belonging'.

This is undoubtedly a new circumstance, an appropriate response to change and to constant in-migration and out-migration. For unlike Whalsay and Elmdon, the Allendales have historically attracted a very transient population of 'incomers' which has ebbed and flowed in accordance with the state of its primary economic resource - lead mining. Hence, the primacy of value accorded to place, as a symbol of community, is well rooted in historical circumstance, transcending the value accorded to residence. Furthermore, it may even be suggested that, as the old 'authentic' Allendonian families become increasingly outnumbered by incomers, or increasingly dissipated through out-migration, place in fact increases its value as a source of reference. Respect for place, the strength of attachment to place, and observation of the traditional rules and norms of place will perhaps become more important as idioms for distinguishing between groups, and for regulating and classifying their affiliations and their degree of 'belonging'.

REFERENCES

Bourdieu, P., 1971. The Berber House or the world reversed, in: *Exchanges et Communications: Melanges Offerts a Claude Levi-Strauss a l'Occasion de son 60 Anniversaire,* (Mouton), pp. 151–161, 165–169.

Cohen, A.P., 1982. A sense of time, a sense of place: the meaning of close social association in Whalsay, Shetland, in: A.P. Cohen (ed.) *Belonging,* (Manchester: Manchester University Press).

Cohen, A.P. (ed.), 1982. *Belonging,* (Manchester: Manchester University Press).

Cunningham, C., 1973. Order in the Atoni House, in: R. Needham (ed.) *Right and Left,* (London: University of Chicago Press).

Douglas, M., 1970. *Purity and Danger,* (London: Penguin).

Emmett, I., 1982. Place, community and bilingualism in Blaenau Ffestiniog, in: A.P. Cohen (ed.) *Belonging,* (Manchester: Manchester University Press).

Hocart, A.M., 1970. *Kings and Councillors,* (London: University of Chicago Press).

Levi-Strauss, C., 1969. *Totemism,* (London: Penguin).

Levi-Strauss, C., 1972a. *Structural Anthropology* (London: Penguin).

Levi-Strauss, C., 1972b. *The Savage Mind,* (London: Weidenfeld & Nicholson).

Mewett, P.G., 1982. Exiles, nicknames, social identities and the production of local consciousness in a Lewis crofting community, in: A.P. Cohen (ed.) *Belonging,* (Manchester: Manchester University Press).

Quale, B., 1982. Participant observation research in rural Northumberland, *North East Local Studies no. I* (University of Durham).

Quale, B., 1982. Allenheads lives, *New Society,* 61, (1036).

Quale, B. and Hockey, J., 1981. Keeping the Dale fires burning, *New Society,* 55, (946).

Strathern, M., 1982. The village as an idea: constructs of villageness in Elmdon, Essex, in: A.P. Cohen (ed.) *Belonging,* (Manchester: Manchester University Press).

14. State planning and local needs

IAN GILDER

The phrase 'state planning and local needs' implies a coherence to the activities of central and local government in rural areas, which I would argue does not exist. The principal aim of this paper is to explore the reasons for this fragmentation of policy and in so doing to suggest ways of altering and improving the response of governments, both central and local, to rural problems. The satisfaction of local needs is just one facet of rural policy, but at the same time one that highlights the difficulties inherent in creating coherent rural policies. The paper falls into four parts, dealing with: the genesis of present policies which attempt to deal with local needs in rural areas; the nature of those policies and the results of their implementation; the reasons for the failure of local needs policies; and ways of altering the mechanism of public policy-making to improve the response to specifically rural and local problems.

The increasing centralisation of public policy-making since 1945 has, if anything, progressively reduced the specifically rural components of government policy. Even the agricultural policies of successive governments have lacked an explicit rural dimension. When pressed to take action on rural issues (and these pressures have been weak), governments have responded that specific rural policies should be the preserve of local authorities who, presumably on the basis of intimate knowledge, should be able to take account of local circumstances, whether urban or rural.

Land-use planning is one means of implementing local and national policies within a locality. Indeed, within local government the statutory system of structure and local plans, established by the 1968 Town and Country Planning Act, has a unique role. There is no statutory requirement to prepare plans and policies for other services: formal corporate plans do not have to be prepared. Moreover, development plans act as the only formal linkages for the mediation of central government policies and offer a route for giving a spatial perspective to essentially aspatial policies. Finally they offer the only statutory framework

for co-ordinating the policy decisions and expenditure
priorities of the various agencies influencing rural change.

Many analysts would agree with Shaw (1980) that the
failure of statutory development plans to achieve their
potential has been one of implementation. I would suggest
that this is not the principal reason for their failure.
They are fundamentally unsound vehicles for rural policy.
Nowhere is this more apparent than in their policies for
local needs.

SETTLEMENT POLICIES AND THE CONCEPT OF LOCAL NEED

Rural policies in county structure plans have essentially
taken the form of rural settlement policies. Whether in
areas of rapid growth, continuing change or massive decline,
the principal aims of these policies have been to conserve
the countryside, limit losses of agricultural land to
development and secure changes in the patterns of settlement
which (it is hoped) will increase the range of social,
commercial and public services and enhance the educational
and employment opportunities of rural residents.

The conventional wisdom has been that this is best
achieved by the use of key settlement policies. As Cloke
(1979) clearly expressed, "key settlement policies were
(and still are) seen by most planners as a kind of universal
elixir for the multiple ailments of rural Britain". Key
settlement policies remain the principal rural policies in
almost all the structure plans for rural counties. A
survey by the National Council for Voluntary Organisations
suggests that only three county structure plans (those for
North Yorkshire, Cumbria and Gloucestershire) do not rely on
key settlement policies (Derounian 1980).

The principal argument for the concentration of housing
development in key settlements has rested on the costs of
service provision. It runs thus - "considerable savings
in public service costs may be made if new development in
rural areas is concentrated in a limited number of larger
settlements rather than dispersed throughout the countryside". It is the basis not only for concentrating development within rural areas in key villages but also for
diverting development to towns rather than villages. In
areas of continued depopulation key settlement policies
have sought to promote development in selected villages -
an application of the economic 'growth-centre' theory. In
those rural areas which are subject to extreme growth
pressures these policies have been administratively convenient ways of implementing restraint policies.

The economic rationale for key settlement policies,
however, is exceedingly weak (Gilder 1979). As our work in
West Suffolk showed, though there are undoubted internal
economies of scale in certain fixed services (e.g. schools),

these only apply to a few services and when the cost curves
for all rural services are amalgamated there is no clear-
cut relationship between the costs of services and the size
of settlement. The conventional view that there are distinct
population thresholds for particular services is a gross
simplification. Our initial work in West Suffolk was unable
to elaborate another major issue: namely, are rural residents' expectations of the range and standard of their
services sufficiently low that in overall terms rural
services are less costly than urban services?

Even without investigating this last issue in detail,
all the economic arguments seem to favour the maintenance
and development of the present pattern of dispersed settlements. The lower the rate of growth the stronger become
these arguments. No useful savings can be made by concentrating development in a few large villages. It was hoped
that these initial conclusions would be investigated
further by the Department of the Environment funded review
of rural settlement policies carried out by Martin Voorhees
and Associates (1980). While they do conclude that rural
settlement policies should be derived from detailed local
study, they do not advise how this should be done nor do
they carry the economic argument any further forward.

Despite growing criticism, the predominance of key
settlement policies in rural structure plans persists, and
by virtue of the formal relationship between structure and
local plans they form the implicit basis to rural policies
in most local plans. The widespread adoption of key settlement policies leads directly to planning authorities
designating large areas in which development would only be
permitted to meet 'local needs'.

Local needs policies in structure plans have been only
one means by which local government has tried to reconcile
the conflicting objectives of its intervention in rural
housing. Significantly, the other policy options, such as
the direct provision of council housing, the encouragement
of housing associations and other less conventional solutions, are the responsibility of District Councils. The
two conflicting objectives for government intervention in
rural housing are those of efficiency and equity. On efficiency grounds one might justify intervention to restrict
development in rural areas - because the market fails to
take account of amenity and landscape externalities.
However such an intervention will have regressive distributional consequences which may be opposed on grounds of
equity. It is in the attempt to offset these regressive
distributional consequences of such planning restrictions
that lead to 'local needs' policies being defined.

But, what are local needs? Can they be quantified or
assessed? Most rural structure plans contain references
to local needs and the only thing that is clear from studying them, is that the county planning authorities involved
have only a vague idea of what they are trying to do. The

report *Rural Recovery: Strategy for Survival* by the Association of District Councils (1978) gives a typical example
of the prevalent platitudes:

> It is essential that policies take account
> of local circumstances and needs, and
> the wishes and aspirations of the mix
> of people who live in the country.

The conception of local need used in one structure
plan bears little resemblance to that in another. This
difference does not in most cases result from any manifest
differences between those local needs. In many of the plans
the meaning of the word 'local' is taken as self-evident.
Others have attempted to define local as being a single
village or group of villages or even the whole area to which
the policy applies. In the case of the term 'needs' the
same vagueness applies; in some plans it remains completely
undefined. Others approach the problem by listing specific
instances of need, e.g. housing needs resulting from the
formation of additional households from within the resident
population; housing required for those who work, intend to
work, or worked before retirement, in the specified area.
The first of these instances is often extended into a supply
concept of need to allow sharing or overcrowding to be
eliminated or the removal of substandard housing. In a
number of cases structure plans have confused the concepts
of need and demand.

Much of the recent controversy about the problems
facing people in rural areas centres around the concept of
need. The difficulties and problems surrounding definitions
of need are well known. The summary diagram (Figure 14.1)
is based on Bannister (1980). It presents a four-way
classification of need, following that used by Bradshaw
(1972). Need may be treated as being divisible into:

> expressed need (= demand);
> felt need;
> normative need;
> comparative need - the starting point for the concept
>   of relative deprivation.

The diagram includes examples of the standard methods
usually used to measure each of these forms of need with
respect to housing. The inadequacy of some of the methods
is acknowledged. In particular, local authority housing
waiting lists are very partial in their coverage of housing
needs (Rogers 1976).

LOCAL HOUSING NEEDS

Normative need is generally accepted by housing economists
as the appropriate basis of policy. Needleman (1965)
defines it as "the extent to which the quantity and quality
of existing accommodation falls short of that required to

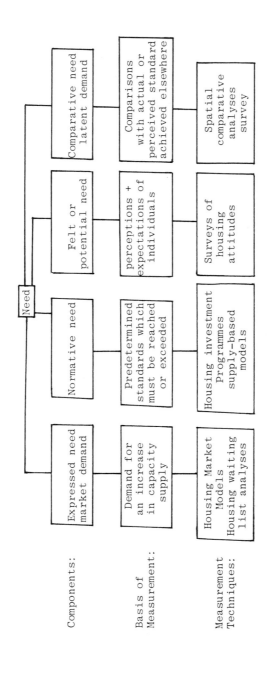

Components:

Basis of Measurement:

Measurement Techniques:

Need

Expressed need market demand — Demand for an increase in capacity supply — Housing Market Models Housing waiting list analyses

Normative need — Predetermined standards which must be reached or exceeded — Housing investment Programmes supply-based models

Felt or potential need — perceptions + expectations of individuals — Surveys of housing attitudes

Comparative need latent demand — Comparisons with actual or perceived standard achieved elsewhere — Spatial comparative analyses survey

Figure   Components of Housing Need (after Bannister 1980)

provide each household with accommodation of a specified minimum standard or above". Housing Investment Programmes (HIP), however, present only a rough guide to the extent of normative needs, even though their assessments of the balance between supply and demand are given considerable weight by the Department of the Environment when allocating housing finance. Their projections of demand, using a formula laid down by the DoE, are essentially demographic and supply based. This approach to housing needs is inadequate for a number of reasons:

1. Single statements of building targets fail to take account of the complex workings of the housing market.

2. Other aspects of needs other than their own narrow definition of normative need are not considered.

3. Each District Council area is treated separately and no account is taken either of housing markets which transcend local authority boundaries or of policies adopted by neighbouring authorities.

4. The 'single number' approach of the HIP submissions do not allow for markedly different housing situations in the same district - as, for example, between rural and urban areas.

The most crucial failure of the HIP approach has been in those areas where local needs policies have been thought necessary. There, they fail to take account of the strength of demand for private housing where supply is restricted - which leads to people from outside the locality being able to outbid local people in the housing market. Restrictive development policies in structure plans tend to intensify competition between the local inhabitants and prospective migrants. This in turn leads to further pressures on the local planning authorities to adopt local needs policies, which as Rawson and Rogers (1976) comment, "tend to be vague and rather bland in their approach, strong on statements of intent but weak on mechanism".

Perhaps the best documented attempts to impose a 'local needs' policy have been those by the Lake District and Peak National Park Boards to use the mechanisms of Section 52 agreements under the 1971 Town and Country Planning Act. The frustrations felt by the Lake District Special Planning Board which led to the adoption of this policy (and its misconceptions of the housing problems of the Lake District) have been well described by Shucksmith (1981). To summarise, the policy of using Section 52 agreements was adopted in 1977 in the draft National Park Plan. The principles are simple: "all new permissions for residential development inside the Park would only be granted if applicants were to sign an agreement which defines the area in which the initial and all subsequent occupiers can work". This policy is very similar to the imposition of planning conditions on rural housing developments to restrict occupation to those employed in agriculture or forestry - the so-called 'agricultural worker conditions', which are in widespread use by rural planning authorities.

The Lake District Board's Section 52 policy has been subject to widespread and telling criticism:

(i)   There are continuing doubts about its legal validity. So far it has not been tested on appeal or in the courts. The DoE maintains that the policy is invalid and deleted similar policies from the Cumbria Structure Plan (which was subject to their approval where the National Park Plan was not). The DoE has consistently argued that planning applications should be decided on the merits of the proposed use, not that of the proposed user.

(ii)   Covenants imposed as a result of Section 52 agreements can be overturned by a decision of the Lands Tribunal. In any case, the only remedy open to the Park Board in the event of a breach of an agreement would be to seek damages.

(iii)   Perhaps most significantly, agreements affect the prices of new, and hence, existing housing. There is evidence that the use of these agreements has reduced the price of those new houses to which they have been applied. Even so, few of these dwellings are at a suitable price for local buyers. Application of the policy has, at the same time, trans-ferred demand to the existing housing stock forcing up the price of those houses. Local people will, in general, wish to participate in the market for existing housing.

(iv)   The policy is inequitable in that it confers enhanced values on some property (and hence owners) at the expense of others.

(v)   Building Societies have refused to give mortgages on houses subject to the Section 52 agreements.

(vi)   Adoption of the policy has allowed the Lake District Planning Board to reduce considerably the number of planning consents issued. This is putting further pressure on house prices throughout the Lake District.

Local needs policies in structure plans, and their develop-ment through policies in the Lake District, have not been working and it is somewhat surprising that such ill-conceived policies could have been applied with any hope that they would work.

The second route by which local authorities, in this instance the District Councils, have been attempting to meet local needs for housing has been by means of their own council housing. However, the provision of rural council housing has been consistently at a low level and there is considerable evidence that housing authorities have not been meeting even the expressed need for housing in rural areas (as measured by their waiting lists). Shucksmith (1981) presents some statistics that suggest that, between 1968 and 1973, urban authorities were building an average of 1 council house for every 1.1 private houses, whilst rural authorities were building 1 house for every 4.5 private houses. Newby (1980) saw this as a matter of political control in the former rural District Councils, dominated by the agricultural interests whose major concern was to minimise rate levies.

What has happened since 1974? The East Anglian authorities examined by Newby and his colleagues in the early 1970s were reconstituted as mixed urban and rural authorities. The agricultural influence on these councils has substantially diminished to be replaced by urban, albeit Tory, interests. Overall housing provision by the new District Councils has been limited by their continued desire to maintain low rates and by a steady reduction in HIP allocations. The majority of the small number of new houses built by the District Councils have been located in the urban parts of their areas. Take one authority, St. Edmundsbury, as an example. Since 1976 the council has built only one dwelling for every ten private dwellings completed in the rural areas, while achieving nearly double that ratio in the urban areas.

In part this pattern of building has been the result of a belief, among both members and officers, that rural applicants would prefer to be housed in nearby towns, to which many of them already travel to work. The evidence for this belief is slight. Another minor influence on the pattern of housing provision has been the higher unit costs of rural housing schemes, which I would estimate to be around 20 per cent. Higher unit costs are a product, in part, of restraint policies in county structure plans which have resulted in enhanced land prices in areas experiencing substantial growth pressures.

Since 1977, central government has controlled the housing activities of District Councils by the Housing Investment Programme procedure. Overall levels of housing allocations have been substantially reduced by the government in accordance with its general philosophy towards public expenditure and private housing. There has been a strong bias in allocations towards the larger urban areas. The General Needs Indicators, used by the Department of the Environment to allocate the majority of money between regions, pays most attentions to the supply-demand balance and to such indicators as the percentages of dwellings that are unfit. These indicators are generally representative of urban rather than rural housing problems. General Needs Indicators are now used to determine only 40 per cent of allocations within regions to individual districts, but given the low overall levels of allocation to regions such as East Anglia this gives a somewhat mythical freedom to redistribute resources.

The sale of council houses to sitting tenants, now a duty under the Housing Act 1980, is exacerbating the problems caused by the low levels of District Council new buildings in many rural areas. Although over 120 District Councils applied to have parts of their areas designated 'rural areas' under the Act (which would allow the councils to refuse sales to sitting tenants) few of these applications were granted. The Secretary of State has decided that the only valid criterion for designating 'rural areas' is where second homes are a significant problem.

Other solutions to local housing needs in rural areas - such as the use of housing associations, equity sharing and municipalisation - have also achieved little success. Many District Councils include statements of support for housing associations in their Housing Investment Programmes. Many housing managers, however, have only begrudgingly accepted that housing associations may be able to take over some of the District Council's role as provider of housing and they still harbour doubts about their effectiveness. This luke-warm attitude is probably of less significance in urban areas, where the national and regional housing associations will put their own pressure on District Councils, or will approach the Housing Corporation for direct funding. There are very few housing associations at present active in rural areas and associations with a specifically rural bias are unlikely to be set up without leadership from the District Councils.

THE FAILINGS OF LOCAL NEEDS POLICIES

The failure to satisfy local needs for housing in rural areas has parallels in other public services. The patterns of service change and withdrawal have been extensively discussed elsewhere. Whilst the causes of these changes have to some extent been economic and social forces beyond the control of public authorities there is little doubt that the prevailing wisdom of the existence of economies of scale in public services has also promoted the rundown. Key settlement policies were intended to stem the withdrawal of services from rural areas, but they have not developed as rural service centres, as was expected. The economic perceptions of administrators have led to decisions to relocate services in centres far larger than most key villages.

Only in the case of employment are significant changes of approach taking place. Those planning authorities who sought to restrain employment development to desirable craft industry in key villages are moving or have moved their positions. The view is becoming prevalent among local politicians in the districts and counties that planning policies should not restrict the possibilities of job creation. Propaganda by central government and employers groups has encouraged this change of emphasis. A wide range of initiatives are now being tried by local authorities, including: fully serviced industrial estates; advance factories; mortgages and loans; reduced rent arrangements; the formation of enterprise trusts; and the active encouragement of such bodies as the Council for Small Industries in Rural Areas.

The major reason for the failure of policies in other sectors has been that neither central government nor local authorities nor the *ad-hoc* agencies have any clear idea of their objectives for rural areas. Few authorities have attempted to answer the fundamental question 'do we want rural settlements at all?' The Countryside Review Committee (1977) put forward three reasons for the continued survival of rural settlements: to ensure efficient agricultural production; to conserve the countryside; and to maintain the countryside as a recreational resource for urban residents and tourists. This approach can be described as that of the 'park-keeper'. It fails to account for two other important reasons for the continued support of rural settlements: namely, that villages represent substantial investments in terms of housing and services; and that many people wish to live in rural areas.

If for these or other reasons we decide that we want rural settlements to survive, who should or will be living in them? Elements of four philosophies have been implicit in rural policies:

(i)   A wish to return village communities to some former idyllic situation. This is essentially Pahl's (1970) 'village of the mind'. Planners have suffered as badly from this myth as have many of the incomers to villages.

(ii) A wish to arrest the changes in village communities and to maintain them as they are. Throughout much of lowland Britain this would mean leaving them with minimal local employment and a swollen population of commuters to nearby towns.

(iii) An acceptance of the inevitability of change linked to a concern to ease the transition of rural communities to the next stage in their 'natural' evolution.

(iv) A search for some new vision of rural communities as a blueprint for change.

Most rural settlement policies have derived from an amalgam of these four philosophies, but all show a weak grasp of

both the economic and social realities of rural areas. All
four are dependent on a view of villages as distinctive
social entities, upon some concept of community.

Does recent research throw any illumination on the
simplified views of community employed by public policy-
makers? I would suggest that the following pointers may
be significant. First, while there has been escalating
social and physical mobility into and between rural areas
there has been a growing awareness of localism. As Cohen
(1982) has argued "the more complete grows the power at the
centre, the more vulnerable the periphery becomes, expressing
its anxiety in a localism which stresses distinctiveness
of character". The profound sense of belonging found in
peripheral communities has been magnified by their enforced
contact with wider society.

Second, communities are the conscious creation of their
members. Such collective identities play down individual
differences between their members. Even in lowland Britain
the village (rather than any wider area) remains the object
of localism. The different groups of people inhabiting a
village often have their own distinctive images of the
village each of which has its own reality.

Third, there is ambiguity about many of the constructs
used to order our analysis of village societies. The
village can be both an open and closed construct; class is
either a social or economic classification. Describing
the linkages in rural society as being a matter of class or
kinship or residence is dependent on the preconceived
notions of the analyst concerned.

These views go some way towards supporting the idea
that there must be a specifically rural and local dimension
to public policy. It is not going to be possible to create
a consensual view of rural communities: policy as a result
must be generated locally. Social change in rural communities has been less than was once thought and this gives
added weight to the third of the approaches I described -
that policies should aim to smooth out the process of
adaption in rural areas. A direct corollary of the failure
to clarify objectives has been a failure to deal with rural
problems in a comprehensive manner. This is due to the
split of functions between national and local government
and the ad-hoc agencies; and to the failure of both central
and local government to establish effective corporate
management, with the consequent dependence of local government on physical planning policies to achieve social policy
objectives.

Plentiful research effort has been given to cataloguing
problems facing individual services. Much of this research
has been followed up by detailed local studies. This
'arithmetic of woe' as McLaughlin (1981) described it has
led almost naturally to attempts to identify rural deprivation. It is only in this last stage that any attempt is

being made to take a comprehensive view. This recent research effort is unlikely to be translated into comprehensive
policy unless we can tackle the split of functional responsibility and the problems of corporate policy in local
government.

Many observers feel that we shall never achieve rational
comprehensive policy-making in local or central government.
Stewart (1982) has recently argued that while the rational
model may be a good way of justifying policy it will never
apply to the creation of policy. A critical element of the
rational model is that of measuring output. Few politicians
or officers are sufficiently disinterested to encourage
serious attempts to measure outputs. By the same token,
few politicians or officers are prepared to articulate the
distributional effects of the policies they advocate
(Webster 1982). The issues are in many cases too complex
for individuals to handle. Even if a rational policy
framework can be devised there remains the problems of
implementation. Before turning into delivered services,
policies travel through a hierarchy of filters, which
include departmental rules and conventions: departmental
organisation, both spatial and hierarchical; and the discretion of those administering the service.

The final reason for the failure to create effective
public policy to meet the social needs of rural areas has
been the lack of political strength of rural interest groups.
Whilst agricultural and landowning interests have been very
successful in pressing for policies to meet their wishes it
is only recently that, through such organisations as Rural
Voice, the social and economic interests of the ordinary
rural dweller are being heard. Neither of the two main
political parties sees rural interests as especially
important and it is probably only through effective pressure
groups that government will ever be provoked to give attention to specifically rural issues.

NEW DIRECTIONS

Pressure must be brought to bear in three specific directions
if there are to be improvements in our system of rural
policy making: to redefine the responsibilities of central
and local government with reference to rural policy; to
work towards corporate management in local government as
the means to create more comprehensive strategies for rural
areas; and to reallocate public resources between urban and
rural areas. These aims are of course interlinked. We
are moving, albeit slowly, towards an acceptance of the
concept of a regional government in Britain. Pressure for
this change must be intensified, bringing with it a system
of local government based on the District rather than the
County Council. Before implementing this change the issue
of accountability has to be resolved. An effective system
of local taxation and of redistribution of tax revenues is

required. Elected regional authorities could then take over the *ad-hoc* agencies (such as the Water Authorities).

Rural policies should become predominantly the responsibility of the new District Councils. The model for their activities in rural areas would be the remarkably successful development agencies such as the Highlands and Islands Development Board. For such a reformed local government to have even a chance of success, central government interference would have to be much reduced. Many of the present failings of physical planning in rural areas can be blamed upon the Department of the Environment's search for consistency between policies for different areas and its refusal to allow structure plans to include detailed policies for matters other than land use. In some ways the abandonment of national rules and standards is going to be the most difficult to achieve. As Cohen (1982) says, in a slightly different context, "locality is anathema to the modern political economy".

The redistribution of responsibility between central and local government is not a prerequisite for the adoption of a corporate approach by local authorities but it would give a significant added impetus to it. Any form of comprehensive land-use planning, such as that intended by the development plan system, will only work as part of a wider corporate policy for the provision of local services and intervention in local economies. Comprehensive approaches to rural policy will only evolve slowly. At best the approach will be one of selective rationality. By advocating the rational approach I am not suggesting that consensus in objectives and policies can be achieved. The skill of the policy analyst lies in opening up the policy process at its most vulnerable points and in making explicit the likely results of particular decisions. This calls for detailed research on local needs. It is only by this process of explicit and informed decision-making in a strategic context that appropriate objectives and policies can be formulated for rural areas.

This brings us neatly back to the third reform being suggested - the reallocation of public resources between urban and rural areas. The reform of local government, the creation of a sensible system of local taxation and the creation of comprehensive rural strategies would achieve that reallocation. If that seems too remote a prospect, then changes to the formulae for Block Grant and Housing Investment Programmes would in themselves yield considerable benefit to rural areas. Local needs policies in land use plans will never work. Money in the hands of District Councils to enable them to provide housing and other services is more likely to satisfy local needs in rural areas.

REFERENCES

Association of District Councils, 1978. *Rural Recovery: Strategy for Survival,* (London: ADC).

Bannister, D., 1980. *Transport Mobility and Deprivation in Inter-Urban Areas,* (Farnborough: Saxon House).

Bradshaw, J., 1972. A taxonomy of social need, in: G. McLachlan (ed.) *Problems and Progress in Medical Care,* (Oxford: Oxford University Press).

Cloke, P., 1979. *Key Settlements in Rural Areas,* (London: Methuen).

Cohen, A.P. (ed.), 1982. *Belonging: Identity and Social Organisation in British Rural Cultures,* (Manchester: Manchester University Press).

Countryside Review Committee, 1977. *Rural Communities: A Discussion Paper,* (London: HMSO).

Derounian, J., 1980. The impact of structure plans on rural communities, *The Planner,* 66, p. 87.

Gilder, I.M., 1979. Rural planning policies: an economic appraisal, *Progress in Planning,* 11, (3).

McLaughlin, B.P., 1981. Rural deprivation, *The Planner,* 67, pp. 31-33.

Martin Voorhees and Associates, 1980. *Review of Rural Settlement Policies, 1945-1980,* (Unpublished research study for the DoE).

Moseley, M.J. *et al.*, 1978. *Rural Transport and Accessibility,* (Norwich: University of East Anglia).

Needleman, L., 1965. *The Economics of Housing,* (London: Staples Press).

Newby, H., 1980. *Green and Pleasant Land?,* (Harmondsworth: Penguin).

Pahl, R. E., 1970. Newcomers in town and country, in: R.E. Pahl *Whose City?,* (London: Longman).

Rawson, M. and Rogers, A.W., 1976. *Rural Housing and Structure Plans,* (Countryside Planning Unit, Wye College, Kent).

Rogers, A.W., 1976. Rural housing, in: G.E. Cherry (ed.) *Rural Planning Problems,* (London: Leonard Hill).

Shaw, M.J., 1980. Rural planning in 1980: an overview, *The Planner,* 66, pp. 88-90.

Shucksmith, M., 1981. *No Homes for Locals,* (Farnborough: Gower).

Standing Conference on Rural Community Councils, 1978. *The Decline of Rural Services,* (London: SCRCC).

Stewart, J., 1982. Choice in the design of policy systems, in: S. Leach and J. Stewart (eds. *Approaches in Public Policy,* (London: Allen & Unwin).

Webster, B.A., 1982. The distributional effects of local government services, in: S. Leach and J. Stewart (eds.) *Approaches in Public Policy,* (London: Allen & Unwin).

NOTES ABOUT THE AUTHORS

TONY BRADLEY B.Sc. is lecturer in social policy and Director of distance learning at the Chelmer Institute of Higher Education, Essex. He has done postgraduate and other research into various aspects of rural social change. Most recently he was engaged (with B.P. McLaughlin) on a research project on 'Deprivation in Rural Areas' funded by the Department of the Environment.

GRAHAM COX M.A. is lecturer in sociology at the University of Bath. He is currently engaged, with Philip Lowe and Michael Winter, on research on land-use politics funded by the Economic and Social Research Council.

DIANA FORSYTHE B.A., M.A., Ph.D. is a social anthropologist currently doing field research in Germany. She has held a research post at the University of Aberdeen, and teaching posts at the Universities of Bielefeld, Cologne and Lawrence (Wisconsin). She is a senior associate member of St. Antony's College, Oxford. She is the author of *Urban/Rural Migration, Change and Conflict in an Orkney Island Community* (SSRC 1982) and *The Rural Community and the Small School* (University of Aberdeen Press 1983).

IAN GILDER M.A., Dip. T.P. is a planning officer with St. Edmundsbury Borough Council, Suffolk, responsible for operational research, local planning and information services. His main research interest is on rural settlement policy. He is the author of *Rural Planning Policies: An Economic Appraisal* (Pergamon 1979).

J. HERMAN GILLIGAN B.A., P.G.C.E. did postgraduate research at the University College of Swansea, where he is now employed as administrative secretary of the Institute of Health Care Studies. His research interests include community studies and the social anthropology of Britain.

PHILIP LOWE M.A., M.Sc., M. Phil. is lecturer in countryside planning at University College London. His research interests lie in environmental politics and rural planning. He was co-author of *Environmental Groups in Politics* (Allen & Unwin 1983). He was convenor of the Rural Economy and Society Study Group between 1981 and 1984, and is British co-editor of *Sociologia Ruralis*.

TERRY MARSDEN B.A., Ph.D., formerly tutor in geography and lecturer in social policy at University College Swansea, is now senior lecturer in town planning at the Polytechnic of the South Bank, London. His research interests are in social and economic changes in rural areas and the urban fringe. He became convenor of the Rural Economy and Society Study Group in 1984.

RICHARD MUNTON B.A., Ph.D. is lecturer in geography at University College London. He has conducted considerable research in the fields of land management (primarily in the urban fringe) and land tenure change, and was an adviser to the Northfield Committee of Inquiry into the occupancy and acquisition of agricultural land. He is author of *London's Green Belt: Containment in Practice* (Allen & Unwin 1983) and co-editor with Judith Rees of Allen & Unwin's Resource Management Series.

BRENDAN QUAYLE B.A., Dip. Anth., M.A., Ph.D. is research fellow and tutor in the Department of Adult Education, University of Durham. He has conducted social anthropological research in the rural communities of northern India (1975-80) and northern England (1979-date). He is co-director of the North East Local Studies Group.

GARETH REES B.A., B. Phil. is lecturer in the Department of Sociology, University College, Cardiff and in the Department of Town Planning, University of Wales Institute of Science and Technology, Cardiff. His research interests include the political economy of urban and regional development and Welsh social structure. He was author of *Poverty and Social Inequality in Wales* (Croom Helm 1980), and is currently completing a book on British inner cities.

SUE STEBBING B.Sc., Ph.D. graduated from Wye College, University of London in 1978 with a degree in Rural Environment Studies. She was awarded a Ph.D. in 1982 for research into women's roles in contemporary rural England. She does freelance research on rural social issues in between bringing up two children.

MARILYN STRATHERN M.A., Ph.D. worked for a number of years (1964-76) in Papua New Guinea on gender relations, legal anthropology and urban migration. She has since revived an early interest in English village life through analysis of materials on Elmdon collected by the Department of Social Anthropology, Cambridge University. Her books include *Women in Between* (Seminar Press 1972); *Nature, Culture and Gender,* co-edited with Carol MacCormack (Cambridge University Press 1980); and *Kinship at the Core* (Cambridge University Press 1981).

JOHN URRY M.A. Ph.D. has taught in the Department of Sociology, University of Lancaster since 1970. At present he is senior lecturer and Head of Department. He is the author and co-author of various books including, with R. Keat, *Social Theory as Science* (Routledge & Kegan Paul 1982); *The Anatomy of Capitalist Societies* (Macmillan 1981); with N. Abercrombie, *Capital, Labour and the Middle Classes* (Allen & Unwin 1983); and with other members of the Lancaster Regionalism Group, *Localities, Class and Gender* (Pion 1984). He was co-editor (with D. Gregory) of *Social Relations and Spatial Structures* (Macmillan 1984).

MICHAEL WINTER B.Sc. is a research officer at the University of Bath investigating conservation and agricultural policies. He has a degree in rural environment studies from Wye College,

University of London. He has conducted research on the sociology of agriculture at the Open University and the University of Exeter. He is British co-editor of *Sociologia Ruralis,* and was convenor of the Rural Economy and Society Study Group between 1980 and 1981.